Models and Algorithms for Multi-objective Transportation Optimization Problems

Zhang Mo

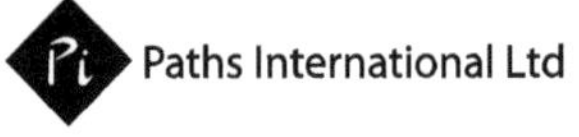

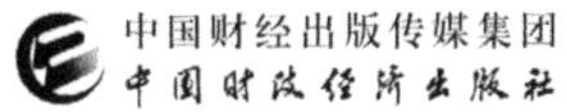

We are grateful to the financial support of National Natural Science Foundation of China (NSFC, #71972099).

前言

本书研究了现代物流行业中被广泛关注的两类优化问题：一类是运输服务采购的优化问题；另一类是线上线下（O2O）外卖配送的优化问题。

在全球化背景下，运输服务在全球贸易的地位越来越关键，并且运输服务的外包已然成为现代物流业中的普适性手段。全球贸易公司（货主）通常采取反向拍卖的形式从全球运输服务商（承运人）处购买运输服务，以满足自己的货运需求。运输服务采购方案的优化可以帮助货主在运输计划阶段节省成本。除此以外，大多数货主还关心除成本因素以外的其他因素，例如服务质量 和服务可靠性等。

首先，本书研究了运输服务采购中的双目标优化问题，该问题是对经典运 输服务采购问题的扩展，即以最小运输成本和最短运输时间为目标。对于货主而言，即使需要支付额外的运输成本，较短的运输时间也更具吸引力，因为敏捷的 供应链可以更好地帮助货主增加收入。相对于海运，公路卡车货运的运输服务商的数目通常是几十个，精确方法无法找到所有非支配解。本书提出了基于 TPEA 算法的两阶段多目标优化算法。仿真结果表明，本书提出的算法优于其他两种精确方法和经典的启发式方法 NSGA-II，并且能够得到大多数实际应用规模算例的非支配解集。

其次，本书研究了另外一种同时考虑运输时间和总量折扣的双目标运输服务采购的优化问题，提出了一种新的双目标分支定界算法，该算法采用新的上界集和两个更强的剪枝规则来提高计算帕累托前沿的效率，同时采用两种有效缩小搜索区域的方法来改进下界集，还提出了一种混合分支方法（先在决策空间分支后再在目标空间中分支）。计算实验生成了一组模拟满载集装箱海运服务采购的测试数据。仿真结果表明，本算法能够解决实际规模的算例。进一步分析表明，将总量折扣引入运输服务采购可以有效地改善货主的决策。

最后，本书研究了面向 O2O 平台的带时间窗的双目标外卖配送路径优化问题，以最小化骑手的总工作时间和客户的总等待时间为优化目标。对于大多数 O2O 外卖平台，除了通常考虑的日常物流运营成本之外，还需要考虑为客户提供的服务质量等因素，因为它可能也会影响该平台在行业的竞争力。本书构建出一个双目标模因算法，它在多目标进化框架中加入多方向局部搜索策略，并使用了路径组合方法进一步提升最终的帕累托前沿。实验结果表明，该算法能够有效改善 O2O 应用中的外卖平台决策。

Preface

This book studies two classes of decision-making optimization problems widely concerned in the modern logistics industry: transportation service procurement problems and online-to-offline(O2O) food pickup and delivery problems.

With the rapid development of trade globalization, transportation plays a critical role in global trade, and outsourcing transportation has become a very common practice in the modern logistics industry. A global company (shipper) often uses reverse auctions to purchase transportation service from global transportation service providers (carriers) to meet its shipping requirements. Optimized transportation service procurement solutions help shippers save transportation costs before any freight is moved in transportation operations. In addition to transportation costs, other factors such as service quality and service reliability are also very concerned by most shippers.

The first problem studied in the book is the full truckload transportation service procurement with transit time (FTL-TPTT). It is an extension of the classic transportation service procurement problem by minimizing the total transportation cost and the total transit time simultaneously. A solution with a shorter transit time is more attractive for a shipper even if additional transportation cost is required, since agile supply chains can better help the shipper to improve its revenue. Compared to sea shipping, the number of carriers for trucking freight transportation is often dozens, and multi-objective exact methods are not capable of finding all the nondominated solutions. Therefore, we design a two-phase evolutionary algorithm (TPEA) to solve the FTL-TPTT. Computational results show that the bi-objective heuristic outperforms two multi-objective exact methods and the classic NSGA-II, and it is able to solve most of the practical size instances optimally.

The second problem is the bi-objective transportation service procurement with transit time and total quantity discounts (TPTT-TQD). We develop a bi-objective branch- and-bound algorithm. Two upper bound sets are designed and two strong fathoming rules are proposed to enhance the efficiency of computing the Pareto front. Two bounding methods with search region reduction are developed to improve the lower bound sets. Moreover, a hybrid branching method is also proposed; it firstly branches in the decision space and then branches in the objective space. A set of test instances that simulates the full container load sea freight service procurement is generated. Computational results on the test instances show that the algorithm is capable of solving the instances of practical sizes. Further analysis shows that decisions can be improved if total quantity discounts are introduced in the procurement of transportation services.

Finally, the book studies the bi-objective O2O food pickup and delivery problem with time window (O2O-PDPTW), which minimizes the total duration time of all the riders and the total waiting time of all the customers simultaneously. For most O2O takeout platforms, in addition to daily logistics operating costs typically considered, the service level needs to be considered as it may be also likely to affect the competitiveness of the platform in the takeout industry. A bi-objective memetic algorithm is proposed. It embeds a multi-directional local search in the multi-objective evolutionary framework to search the nondominated solutions. The algorithm also uses a route combination method to further enhance the final Pareto front. Experiments show that the algorithm can help O2O takeout platforms improve decisions.

Contents

Chapter 1

Chapter 2

Chapter 5

List of Tables

List of Figures

Chapter 1
Introduction

1.1 Background

Effective transportation of goods is critical to the success of most manufacturing and retail companies. In the transportation market, companies that need to move freight are called shippers. Many shippers now outsource transportation and logistics in their value chains to more dedicated carriers in the transportation industry. Especially, for some global shippers, the annual transportation cost of moving raw materials and manufactured products could be multi-billion dollars. With the incentives to reduce costs, centralized procurement activities for purchasing transportation service from carriers are held every year. Auctions are usually conducted by a shipper to purchase transportation services. Optimized transportation service procurement solutions help shippers save transportation costs before any freight is moved in transportation operations.

Furthermore, with the rapid development of Internet technology and e-commerce logistics, some companies also outsource transportation and logistics to online-to-offline (O2O) platforms, which provides transportation services as its offline key business. Nowadays, O2O e-commerce becomes more and more popular. O2O platforms allow customers to order goods (service) online, and the goods (service) will be delivered offline. O2O e-commerce business model combines the advantages of online and offline to make the supply chain faster, thus increasing customer satisfaction and bringing more profits to the company. In the O2O takeout market, O2O takeout companies, such as Meituan and Ele.me, provide platforms to match food supplies and demands between restaurants and customers. Once an online order for some food of a restaurant is received from a customer, the O2O takeout company will assign the order to a rider to fulfill the offline pickup and delivery service; that is, the rider will pick up the food from the restaurant and deliver it to the customer. In China, the market scale of O2O takeout industry is increasing with remarkable speed and is expected to exceed 3600 billion in 2018[92]. More than 1/5 of total population in China has already became the users of O2O takeout market [56]. In practice, an O2O takeout platform usually runs an optimization engine to seek solutions that allocate orders to riders and determine the delivery routing plans to serve all the orders so as to minimize the operating costs. Such an optimization problem computed by the O2O platform is an O2O food pickup and delivery problem.

For most real-life transportation problems, decision makers are confronted with multiple conflicting objectives. In general, there is usually a trade-off between minimizing costs and maximizing service levels. In addition, other factors, such as reliability and availability, can also drive the decision making. In most practical situations, two or more objectives should be considered. These objectives are of almost equal importance to most decision makers. Such problems are well-known as multi-objective optimization problems. Multi-objective optimization problems require more computing power than a single objective, but provide decision makers with a broader perspective and multiple solution options.

The thesis studies the models and algorithms for two classes of separate but related transportation optimization problems considering multiple objectives: multi-objective transportation service procurement problems and multi-objective O2O food pickup and delivery problems. The following studies are then considered.

1.1.1 Transportation service procurement problems

Being a very significant component of the international logistics, transportation service procurement plays an important role in the transportation industry.

For a global shipper, centralized procurement activities for purchasing transportation service from carriers are held every year. At the beginning of each year, the procurement team in the company organizes several regional auctions to decrease the auction size and impose centralization. Different transportation modes (e.g., trucking, air cargo and sea freight) are considered in the auction.

The process of a transportation service procurement auction is depicted in Figure 1.1[40]. It includes three stages: pre-auction, auction and final decision.

- Pre-auction stage: a shipper aggregates freight demands and transportation requests from different business units and organizes a multi-round reverse procurement auction. A set of carriers are invited to participate in the auction. It should be mentioned that only those qualified carriers are invited to bid in the auction.

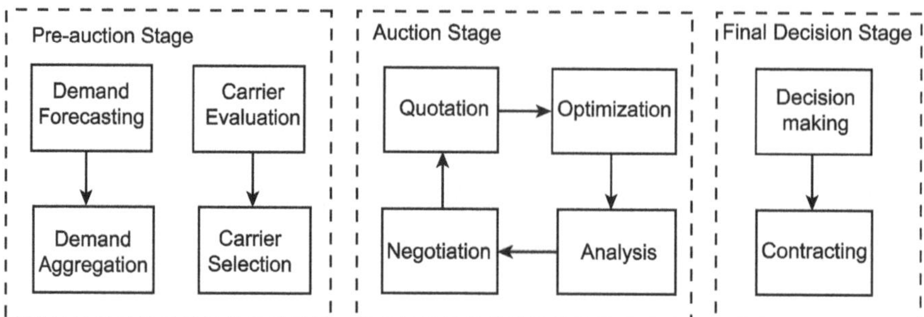

Figure 1.1 Transportation service procurement process

- Auction stage: at each round, the shipper receives bids from the carriers. Every carrier submits a bid for each shipping lane operated by the carrier. In general, a bid contains the shipping rate, capacity of the carrier on the shipping lane and the discount strategy. After receives bids from the carriers, the shipper runs an optimization engine to generate allocation plans. An allocation plan decides a set of winning carriers and allocates the corresponding freight volume. The optimization engine is run multiple times since different scenarios can be constructed for analysis.
- Final decision stage: based on the analytical results, the shipper proposes strategies for negotiations or decides the final allocation plan.

In the procurement, minimum quantity commitments (MQC) [48] and quantity discounts are effective tactics to reduce cost for the shipper and to improve revenue for the carriers. After a round of the auction, the shipper negotiates with some carriers to request lower transportation rates or discounts. Meanwhile, with the aim of improving revenue, a carrier may request more freight volume to be guaranteed by the shipper in exchange for attractive discounts or lower transportation rates. In the next round of the auction, the carriers update their bids with new bidding rates or discounts. The discounts are called the total quantity discounts (TQD) [37]. A total quantity discount is activated only if the total allocated freight volume on all the lanes is equal to or larger than the agreed MQC.

In the traditional procurement, the critical objective in the shipper's decision is to minimize transportation cost. However, as revealed by both literature [18] and practice, the non-price factors such as transit time [40], reliability, service quality [15, 16], and carrier reputation [79] also have significant impacts on the shipper's decision. Following Hu et al. [40], we also consider minimizing the transit time in a separate objective in the optimization model.

1.1.2 O2O food pickup and delivery problems

With the rapid growth of smart phone and the internet, online-to-offline (O2O) e-commerce business model becomes more and more popular. The O2O takeout companies enable customers to purchase food online and receive food offline.

The O2O food pickup and delivery process is described in Figure 1.2. With the O2O takeout platforms, the customers purchase food through mobile APPs. Such a platform matches the food demands from customers and supplies from restaurants, and manages delivery teams to execute the food deliveries. Once a food order is received, the platform of the company will notify the restaurant to prepare food and assign the delivery for the order to a rider. The rider then goes to pickup the food at the restaurant, and deliver it to the customer.

The O2O food pickup and delivery model is different from the classic pickup and delivery model. In the O2O food pickup and delivery, the distinctive feature of rider routes is that they end at the last serviced customer instead of at a depot. Hence, the solution is a set of "open" routes, where the duration of each route is calculated from the depot to the last customer.

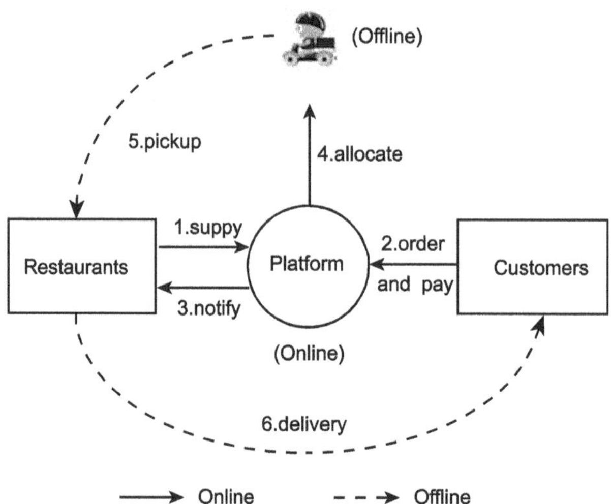

Figure 1.2 O2O food pickup and delivery process

Traditional O2O takeout platforms assign an order to the nearest rider. Unfortunately, such a simple strategy may not be helpful in improving the pickup and delivery solution for the logistics system that deals with all the existing orders. In practice, an O2O takeout platform usually runs an optimization engine to seek solutions that allocate orders to riders and determine the delivery routing plans to serve all the orders so as to minimize the operating costs. In the classical pickup and delivery problems, to minimize the total travel time or travel distance is quite common. However, such a single objective might not be enough for the O2O takeout platform. In the O2O takeout market, customer satisfaction is much more important for the success in the long-term intense competition with rivals.

1.1.3 Multi-objective optimization

A single-objective optimization problem involves a single objective function and have optimal solutions with the same objective value. However, a multi-objective optimization problem considers several conflicting objectives simultaneously, and there is usually no single optimal solution, but a set of solutions with different objective values in the Pareto front.

The thesis presents a basic concepts and problem formulations of *multi-objective optimization problems* as follows:

$$\min\{f_1(x), f_2(x), \cdots, f_k(x)\}$$
$$s.t. x \in X \tag{1.1}$$

The formulate has k objective functions f_i that we aim to minimize simultaneously. The *decision vectors* $x = (x_1, x_2, \cdots, x_k) \in X$ defines the *feasible region* X, also called *decision space*. The image of X in the *objective space* is the set of feasible points denoted by Z.

The thesis focuses on bi-objective optimization problems. The dominance and efficiency in bi-objective optimization are defined as follows.

Definition 1.1. In the objective space, z is said to **dominate** z' if $z_i \leqslant z'_i$ for $i=1, 2$ and either $z_1 < z'_1$ or $z_2 < z'_2$. If $z_i < z'_i$ for $i=1, 2$, z is said to **strictly dominate** z'. z is **nondominated** if there exists no $z' \in Z$ that dominates z. The set of all nondominated points is denoted by Z_N. The thesis sometimes also just uses nondominated solutions to refer to efficient solutions.

Definition 1.2. In the decision space, a feasible solution x is said to be **weakly efficient** if there does not exist any other feasible solution x' such that $z(x')$ strictly dominates $z(x)$, a feasible solution x^* is **efficient** if $z(x^*)$ is nondominated. The set of all efficient solutions is denoted by X_E.

Definition 1.3. Let z_1^* be the minimum value of the first objective. A point $z^{TL} \in Z$ achieving z_1^* is called a *top left* point if there is no other point $z' \in Z$ such that $z'_1 = z_1^{TL}$ and $z'_2 \leqslant z_2^{TL}$. Similarly, a point $z^{BR} \in Z$ achieving z_2^* is called a *bottom right* point if there is no other point z' $\in Z$ such that $z'_2 = z_2^{BR}$ and $z'_1 \leqslant z_1^{BR}$, where z_2^* be the minimum value of the second objective. Both z^{TL} and z^{BR} are called **lexicographic minimal points**.

Moreover, the nondominated points in Z_N are classified as *supported* and *unsupported* nondominated points. Let $conv(Z_N)$ denote the convex hull of set Z_N. The supported nondominated points are on the boundary of $conv(Z_N)$, while the unsupported nondominated points are in the interior of $conv(Z_N)$. The supported nondominated points can be further clustered as **extreme supported nondominated** (ESN) points which are the extreme points of $conv(Z_N)$ and **nonextreme supported nondominated** (NSN) points which are on the boundary but not the extreme points of $conv(Z_N)$. It should be noted that the two lexicographic minimal points z^{TL} and z^{BR} are both ESN points.

1.2 Literature review

In the following, the thesis briefly reviews the literature on transportation service procurement problems, O2O food pickup and delivery problems and bi-objective optimization methods.

1.2.1 Literature on transportation service procurement problems

Caplice and Sheffi[18] first introduced the optimization-based transportation service procurement problem and proposed an optimization model for it. It introduced many practical business considerations concerned by the shipper when the decision was made. Such business considerations were formulated by mathematical terms in the model. Since then, studies on transportation service procurement have been dramatically increased. Transportation service procurement problems and its variants have been extensively studied over the last two decades. The transportation service procurement problems can be divided into two categories: the winner determination problem from the perspective of shippers and the bid generation problem from the perspective of carriers. This thesis concentrates on solving winner determination problems.

Many problems in the field have been studied from the different perspectives of the shipper and the carriers[43,100], under the different auction mechanisms such as combinatorial auction [85], and with different concerns in the decision-making process such as discounts, service quality[15, 16, 79], transit time [40], minimum quantity commitments[47, 48], side constraints [46] and demand uncertainty [80, 99, 103].

In the following, the thesis reviews the literature on transportation service procurement problems with auctions, business constraints and multiple objectives.

1.2.1.1 Auctions and business constraints in the transportation service procurement

Typically, auctions in practice involve the bidding on each lane separately. Recently, combinatorial auctions[85] have drawn a lot of attention for transportation procurement which could lead to more efficient solutions. In a combinatorial auction, a carrier could submit bids on lane packages.

With regard to transportation service procurement problems under uncertainty, Ma et al.[54] first proposed a two-stage stochastic programming method to solve this problem under shipment volumes uncertainty. Lim et al.[48] studied this problem with seasonally varying the shipper demand. Remli and Rekik[80] proposed a two-stage robust formulation to solve the same problem as Ma et al[54].

In the decision and analysis process, the shipper has different business considerations, which can be formulated into constraints and added into the model of the decision-making problem. Caplice and Sheffi [18] summarized a variety of business considerations that are presented below:

- Minimum/maximum number of carriers: no more than or no less than a certain number of carriers can win.
- Favoring of incumbents: incumbent carriers are often favored, especially on service-critical lanes to key customers or time-sensitive plants.
- Back up carrier bids: carriers are required as a back-up choice on some critical shipping lanes.
- Minimum/maximum Coverage: the amount of traffic that a carrier wins across the system is within certain bounds.
- Threshold volumes: a carrier wins any freight that it is of either a certain minimum threshold amount, or they win nothing at all.
- Service requirement for alternates: carriers are required to act as both primaries and alternates over different segments of the system.
- Restricting carriers: certain carriers are restricted from serving certain portions of the network.
- Complete regional coverage: every carrier should be able to cover all lanes from a certain location.

Moreover, some widely used practical constraints in the industry were also proposed, such as MQC (volume guarantee) [47–49], and lane cost balancing constraint [46]. The MQC (volume guarantee) first studied by Lim and Xu [49] requires that a shipper has to allocate a minimum quantity freight to a winning carrier. Lim et al. [46] first considered the lane cost balancing constraint in the context of freight allocation, which requires the allocation must not assign an overly high proportion of freight to the more expensive shipping companies servicing any particular lane.

1.2.1.2 Total quantity discounts in transportation service procurement

Transportation cost has always been the most critical metric in the objective of the optimization-based transportation service procurement. Centralized transportation service procurement by aggregating freight demands from all the business units is a common way for a global shipper to achieve economies of scale and reduce the transportation cost. To explore more cost reduction, strategic shippers apply pricing strategies such as quantity discounts to facilitate the business between the shipper and the carriers.

Goossens et al. [37] studied a basic purchasing problem with total quantity discounts and several variants where different side constraints were considered. The linear programming (LP) relaxation of the problem was reformulated as a network flow model. A branch-and-bound algorithm was proposed to solve the problem exactly. Goossens et al. [37] pointed out that the total cost can probably be reduced by purchasing more freight volume than the demand and then securing larger discounts. Based on this work, Manerba and Mansini [57] studied a variant where a capacity on each product was considered for each supplier. A branch-and-cut algorithm was proposed by the authors.

By extending the purchasing problem, Mansini et al. [58] studied a supplier selection problem with total quantity discounts. The problem is to minimize the sum of procurement cost and truckload shipping cost in the objective. Heuristics based on integer programming were proposed for the problem. More closely related to the transportation service procurement, Qin et al. [77] investigated a freight allocation problem with the all-units quantity discount for a retail distributor, and proposed a heuristic with tabu search to solve the problem. In the problem, the shipper (distributor) needs to decide the freight volume for each carrier and each carrier requests a minimum total freight volume over all lanes.

1.2.1.3 Multi-objectives in transportation service procurement

In addition to the procurement cost, non-price factors (i.e., service level) have also been studied in the transportation procurement problems [85]. Buer and Pankratz [16] pointed out that service quality should be considered in the optimization model of the transportation service procurement problem. Service quality was quantified with small integers and a bi-objective optimization model, which dealt with the minimization of both the total transportation costs and the service quality, was brought up. Rekik and Mellouli [79] considered carrier reputation to indicate service quality. Instead of optimizing the two objectives, the carrier reputation was formulated into a hidden cost. Therefore, only a single objective related to cost was optimized. Buer and Kopfer [15] studied a metaheuristic to minimize the total transportation costs and maximize the total transportation quality simultaneously in a combinatorial auction.

However, quantifying service quality is a problem itself because different shippers would choose different indicators. Hu et al. [40]

studied the transit time in the transportation service procurement problem. Unlike the service quality which could be complex to be quantified, it was pointed out that the transit time could be obtained easily from historical data. To minimize the transportation cost and the transit time as two separate objectives, the authors proposed a bi-objective branch-and-bound method to explore the set of extreme supported nondominated (ESN) solutions on the Pareto front. They also proposed the feasibility checks to speed up the exploration of the branch-and-bound tree. However, this method could only solve small instances with a maximum of 25 carriers. For sea freight full container load (FCL) services, the notable carriers are only a few and the method is thus applicable. However, in the case of FTL freight services, where the number of carriers could be more than 30, executing the bi-objective branch-and-bound algorithm would be very time-consuming; therefore, it is not a wise option. Designing a multi-objective evolutionary algorithm to approach the Pareto front efficiently will be helpful to practical shippers.

1.2.2 Literature on O2O food pickup and delivery problems

In this section, existing studies on the vehicle routing problems for perishable goods are first investigated. Then, recent studies on single objective and multi-objective optimization methods applied to pickup and delivery problems and other related problems are presented.

1.2.2.1 Vehicle routing problems for perishable goods

In recent years, many researches have studied vehicle routing problems for perishable goods in various aspects [22, 39, 55, 86, 97]. Hsu et al. [39] considered the randomness of perishable goods in delivery process. Ma et al. [55] developed a model which combined order selection and perishable product delivery to maximize the profit. A hybrid ant colony algorithm was developed to solve the model. Song et al. [86] investigated a vehicle routing problem that considered refrigerated and general type of vehicles for multi-commodity perishable food goods delivery. Wang and Yu [97] considered different delivery modes and established a perishable product delivery network to minimize the total costs.

Some researches focused on multi-objective VRPs for perishable goods [2, 12, 95, 96]. Bortolini et al. [12] constructed a multi-objective perishable distribution model which considers operating cost, carbon emissions and delivery time. Amorim et al. [2] and Wang et al. [96] proposed new models that minimize the total costs and maximize freshness state of delivered goods. Moreover, Wang et al. [95] developed a multi-objective vehicle scheduling optimization model for perishable products that maximizes customer satisfaction and minimizes total delivery costs.

1.2.2.2 Single objective pickup and delivery problems

A growing body of research has been published on pickup and delivery problems (PDP) and its variants. For a general review to pickup and delivery problems, the readers can refer to the survey papers by Parragh et al. [71] and Berbeglia et al. [8]. According to Berbeglia et al. [8], PDPs are often classified into three groups by the number of origins and destinations: many-to-many, one-to-many-to-one and one-to-one. In many-to-many problems (M-M), any vertex can serve as a source or as a destination for any good. In one-to-many-to-one problems (1-M-1), goods are initially available at the depot and are destined to the customer vertices; in addition, goods available at the customers are destined to the depot. Finally, in one-to-one problems (1-1), paired pickup and delivery services are considered, and each good has a given origin and a given destination. The applications of one-to-one PDPs arise, for example, in courier operations, door-to-door transportation services and dial-a-ride. Specifically, the O2O food pickup and delivery services is also a kind of one-to-one problems. We refer the reader to the survey paper by cordeau et al. [26] for more details on the one-to-one PDPs.

Many variants of the PDP have been studied by researchers, such as the PDP with time windows (PDPTW), PDP with maximum driver working duration, PDP with transfers [59], PDP with LIFO loading [24], PDP with multiple stacks [21], dynamic PDPs, PDP with loading cost [101].

Due to the complexity of PDPs and the instance sizes in real-life applications, most of the literature has focused on heuristic solution methods, although some work on exact methods has been published [4, 32, 52, 78, 81, 82]. A growing body of heuristics and mutaheuristics have been proposed for the PDPs. Nanry and Barnes [65] first developed a metaheuristic for PDPTW, which tested on instances with up to 50 requests. Li and Lim [45] used a hybrid metaheuristic to solve the problem. Ropke and Pisinger [83] developed an adaptive large neighborhood search heuristic to solve the problem. Lu and Dessouky [53] introduced a construction heuristic which incorporates distance increase into evaluation criterion and time window slack reduction as well. Bent and Van Hentenryck [7] developed a two-stage hybrid algorithm for the PDPTW. The first stage used simulated annealing to decrease the number of vehicles needed. The second phase used a large neighborhood search algorithm to reduce total travel cost.

A problem similar to PDP in the literature is the dial-a-ride problem (DARP), the pickup and delivery problem deals with the transportation of goods, while DARP deals with the transportation of people. A detail survey of the DARP is provided by literature [25, 71].

1.2.2.3 Multi-objective pickup and delivery problems

The objective of the classic PDP is either to minimize the total travel distance of vehicles, or to minimize the total travel time of vehicles. Although most PDPs involves multiple conflicting objectives, the multi-objective PDPs did not receive as much attention in the literature as the other VRP classes. To the best of our knowledge, Zhu et al. [106] is the only study that considers total route length, response time and workload simultaneously for PDPs. The model was formulated as a single vehicle 1-M-1 dynamic pickup-and-delivery problem. A multi-objective memetic algorithm, which is a synergy of multi-objective evolutionary algorithm and locality-sensitive hashing

(LSH) based local search, was designed to solve the multi-objective PDP.

Due to the very limited literature on multi-objective PDPs, the thesis studied multi-objective optimization problems that are closely related to PDPs. With regard to vehicle routing problems, Pacheco et al. [68] investigated a bi-objective bus routing problem that trades off service (i.e., the minimization of the longest route) and operational cost (i.e., the minimization of the total distance traveled). They developed a tabu search within the framework of multi-objective Adaptive Memory Programming. A distinctive feature of these bus routes is that they start at the first stop instead of at a depot (or common point of departure). For a general review to multi-objective vehicle routing problem (VRP) problems, see Bérubé et al.[9].

In DARP, user inconvenience must be considered which can be another term in the objective function or by adding maximum user ride time constraints. The objective function in the work of Diana and Dessouky [28] is to minimize a weighted sum of the total distance traveled, the excess ride time over the direct time for all the customers and the total length of the idle times. Another study by Paquette et al. [69] who considers a dial-a-ride model that takes into account routing cost and quality of service. He combined tabu search with a reference point method to construct a set of uncomparable solutions, and used a dynamic weighted-sum objective function to guide the search. Taking into account the trade off between operational efficiency and service quality, Molenbruch et al. [62] developed a multi-directional local search algorithm. Parragh et al. [73] proposed a variable neighborhood search heuristic coupled with path relinking to minimize transportation cost and average ride time simultaneously.

The thesis deals with a bi-objective O2O food pickup and delivery problem with time windows constraints. The two conflicting objectives are the minimization of total rider duration time and total customer waiting time. To the best of our knowledge, no solution methods have been proposed for this problem so far.

1.2.3 Literature on multi-objective optimization methods

The problems discussed in this thesis are all multi-objective (mixed) integer programming problems, three types of solution methods are typically employed. On multi-objective optimization for the multi-objective (mixed) integer programming problem, the popular methods include multi-objective branch-and-bound, scalarization methods and multi-objective heuristics.

1.2.3.1 Literature on multi-objective branch-and-bound methods

A multi-objective branch-and-bound method is a kind of exact methods that need sufficiently time and space. Multi-objective branch-and-bound methods [76] have been successfully applied to a number of problems, such as spanning tree problems [87], traveling salesman problems [41], and facility location problems [36]. The first multi-objective branch-and-bound method for the multi-objective mixed 0-1 integer program was developed by Mavrotas and Diakoulaki [60]. In the method, linear programming relaxations were solved to obtain an ideal point as the lower bound. The concept of bounds in branch-and-bound was extended to the lower and upper bound set in the multi-objective branch-and-bound Ehrgott and Gandibleux [33]. More recently, instead of using a single ideal point, most researchers mainly focused on reducing nodes to be explored using the lower bound set in the bounding part [6, 36, 74, 87, 88, 94]. Different branching strategies and bounding methods were developed to discard part of the objective space during the branch-and-bound exploration. Sourd and Spanjaard [87] developed a branch-and-bound framework for solving multi-objective integer programs. They constructed a set of hyperplanes separating the set of feasible solutions of the subproblem and the set of nondominated solutions to fathom a node. In the work of Stidsen et al. [88], a so-called Pareto branching (PB) was developed to perform a binary branching in the objective space. A slicing method was proposed to partition the objective space into several slices to obtain better upper bounds. Computational results show that the proposed strategies perform better than the two-phase method.

Recently, Parragh and Tricoire [74] introduced a two-stage branching procedure. If a node in the branch-and-bound tree cannot be fathomed by dominance, the corresponding lower bound set is often partially dominated. The search area that may contain nondominated points consists of several continuous parts, and can be explored separately in the objective space by using the second stage branching. Lauth et al. [36] proposed different branch and cut algorithms based on implicitly and explicitly lower bound sets. Cutting plane algorithms were applied to strengthen different parts of the feasible set.

1.2.3.2 Literature on scalarization methods

Scalarization methods provide a set of solutions with quality guarantee. In general, scalarization methods convert a multi-objective optimization problem into a single objective problem. They include the weighting method [61], the ϵ-constraint method [38], the two-phase method [93], the perpendicular search method [20], and the augmented weighted Tchebycheff method [13].

More recently, Boland et al. [11] presented a balanced box method for bi-objective integer programming problems, which is capable of maintaining an evenly distributed set of nondominated points. In a similar way, a triangle splitting method was introduced for bi-objective mixed integer programming problems in the work of Boland et al. [10]. A possible drawback of scalarization methods compared to the multi-objective branch-and-bound method is that a series of single objective integer programming problem have to be solved iteratively to explore the entire Pareto front.

1.2.3.3 Literature on multi-objective heuristics and meta-heuristic methods

To solve large scale multi-objective optimization problems, multi-objective heuristic and meta-heuristic methods to approximate the Pareto-optimal front were reported in the literature, such as the multi-objective evolutionary algorithm based on decomposition (MOEA/D) [104], multi-objective evolutionary algorithms [17, 27, 104], multi-objective simulated annealing algorithms [5], multi-objective ant colony optimization [30, 102] and the multi-directional local search algorithm [91].

One of the most popular multi-objective heuristics is the nondominated sorting genetic algorithm II (NSGA-II) [27]. NSGA-II is mainly characterized by two aspects: the use of nondomination levels to classify solutions and the use of crowding distances to select solutions in the same nondomination level. The nondomination levels and crowding distances are to achieve intensification and diversification in multi-objective optimization, respectively.

The multi-objective heuristic and meta-heuristic methods have been developed for a variety of problems and applications, such as allocation problems [51], dial-a-ride problems [70] and routing problems [75].

1.3 Contributions of the thesis

The thesis develops several dedicated algorithms to solve multi-objective transportation service procurement problems and multi-objective O2O food pickup and delivery problems. These problems were all derived from industrial applications. Our models and algorithms can be integrated into decision support systems to assist shippers and O2O takeout companies to make better decisions. The contributions of this thesis are three folds, which can be summarized as follows:

- A bi-objective optimization method for full truckload service procurement auctions from the perspective of shippers is developed. Although sea freight full container load services procurement with transit time has been successfully tackled by Hu et al. [40], this is the first attempt to develop bi-objective evolutionary algorithm for solving the full truckload services procurement with transit time. It should be noticed that the notable carriers are only a few in the case of sea shipping whereas the number of carriers could be more than 30 in the case of trucking.
- A new bi-objective transportation service procurement problem is studied, where the transportation costs, the transit time and the total quantity discounts are concerned. This new model extends the general cost function to the total quantity discounts. Under a total quantity discount structure, the Pareto front can be even non-smooth. However, the flexibility and complexity of the decision

making are increased for shippers in the procurement auction. To simultaneously minimize the transportation costs and transit time, a Pareto front consisting of nondominated solutions is explored. To the best of our knowledge, no algorithm for the exact solutions of the problem has yet been published. Thus, this work initiates the study of exact algorithms for the bi-objective transportation service procurement problem with transit time and total quantity discounts.

- A new appealing research topic in PDPs is brought up, which focuses on the O2O takeout applications. Although some researches focus on VRPs for perishable food products, little research has been presented that considers perishable food in PDPs. To our knowledge, the O2O food pickup and delivery problems with the waiting time has not been previously studied. It is inapplicable to develop exact methods to solve practical size instances. Therefore, a bi-objective memetic algorithm is developed and can produce a good approximation of the Pareto front for NP-hard problems. We test the approach on a large number of random instances, which facilitate the future researchers. We also provide new some insights in the effect of the choice of time window width on the Pareto front.

In Figure 1.3, we present an overview of formulation and optimization methods for these real application problems in the thesis.

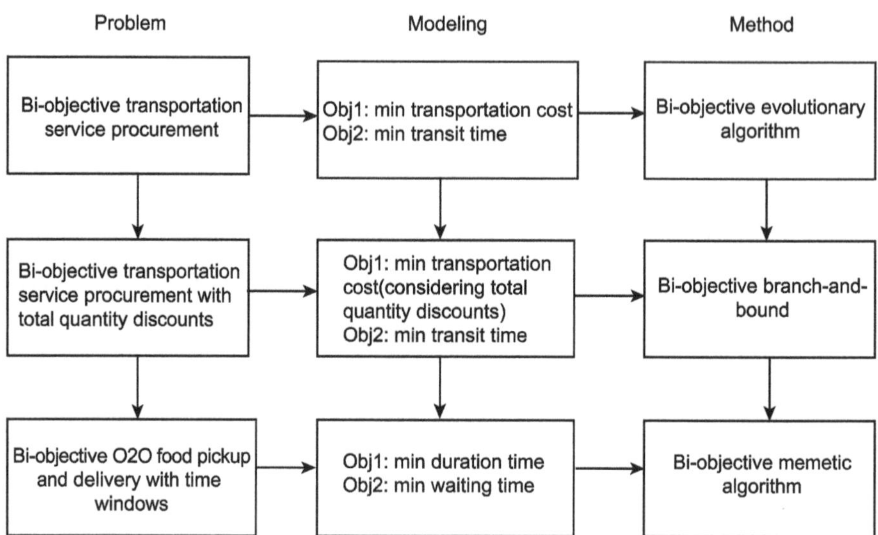

Figure 1.3 Modeling and methods for problems in the thesis

1.4 Organization of the thesis

The remaining parts of this thesis are organized as follows.

- An effective evolutionary algorithm for the bi-objective full truckload transportation service procurement problem (Chapter 2).
  1. Develop a multi-objective evolutionary algorithm to solve the model.
  2. Encode a chromosome using a set of winning carriers to reduce decision space.
  3. Design fitness-based crossover and mutation operators to explore individuals.
  4. Show the algorithm can well approach the Pareto front in computational experiments.
- The bi-objective transportation service procurement problem with transit time and total quantity discounts (Chapter 3).
  1. Develop a bi-objective branch-and-bound algorithm for the transportation service procurement problem with both transit time and total quantity discounts.
  2. Propose two stronger fathoming rules.
  3. Design two novel bounding schemes and a hybrid branching mechanism to reduce search region.
  4. Show the algorithm outperforms the-state-of-art existing methods in computational experiments.
- The O2O food pickup and delivery problem with time windows (Chapter 4).

1. Design a two-phase constructive algorithm that generates an initial set of high quality solutions.
2. Develop a bi-objective memetic method that combines multi-objective evolutionary method with multi-directional local search.
3. Propose accelerating strategies for the bi-objective memetic method.
4. Generate test instances based on data in real applications.
5. Evaluate the performance of the bi-objective memetic method and compare with the state-of-art existing methods.

Finally, Chapter 5 draws the conclusions and discusses possible future directions. The research technology roadmap of the thesis is described in Figure 1.4.

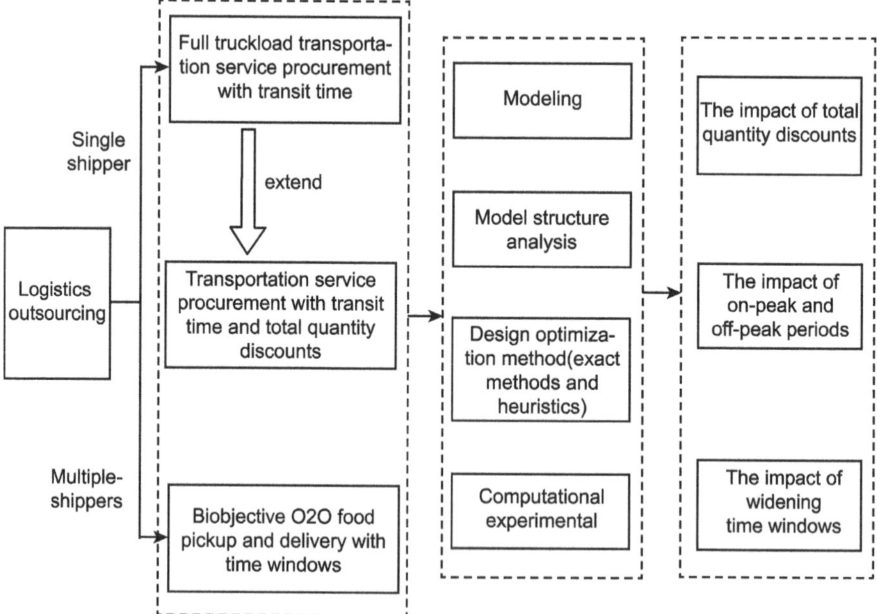

Figure 1.4 Research technology roadmap of the thesis

Chapter 2

Two-phase Evolutionary Algorithm for the Bi-objective Full Truckload Transportation Service Procurement Problem

2.1 Introduction

Auctions are usually conducted by a shipper to purchase transportation services. Caplice and Sheffi [18] described the shipper's decision-making process in a transportation service procurement auction. With all the bids, which include transportation rates and capacities from carriers, the shipper runs an optimization engine to decide a set of winning carriers and allocate freight volume to them on each shipping lane. The optimization engine is usually run multiple times due to scenario analysis and multi-round auctions before the final allocation plan is decided. Therefore, the optimization engine should embed an efficient algorithm for computing good and feasible solutions.

Optimized transportation service procurement solutions help shippers save transportation costs before any freight is moved in transportation operations. In addition to transportation costs, transit time, which indicates service quality and transportation efficiency, is also concerned by many shippers. For a strategic shipper, a solution with a shorter transit time is more attractive even if additional transportation cost is required, since agile supply chains can better help the shipper to improve its revenue. To find the trade-off between transportation costs and transit time, the bi-objective optimization problem in transportation service procurement is to minimize the total transportation costs and the transit time simultaneously.

There are various modes of transportation such as trucking, sea freight and air cargo. This study considers trucking, which occupies a prominent place in the global logistics business. According to the National Bureau of Statistics of China, truck freight accounted for 75.4% of the total logistics market in 2015. For transportation modes other than trucking, the notable carriers are only a few and the exact method is thus applicable. However, in the case of full truckload (FTL) freight services, dozens of carriers can participate in the procurement auction, and executing an exact algorithm would be very time-consuming.

To solve the full truckload transportation service procurement problem with transit time (FTL-TPTT), we design a multi-objective evolutionary algorithm to efficiently approach the Pareto front. In the encoding scheme, the chromosome of an individual is defined by a set of winning carriers. In fact, each individual is a leaf node in a branch-and-bound tree. Instead of implicitly enumerating all the nodes in the branch-and-bound tree as a bi-objective branch-and-bound algorithm does, our algorithm only explores some leaf nodes (individuals) to achieve efficiency. For a leaf node, where a set of winning carriers is determined, extreme supported nondominated (ESN) solutions on the local Pareto front of the node can be explored using a polynomial-time algorithm. The multi-objective evolutionary algorithm is designed in a two-phase framework. The first phase is to explore candidate individuals using the fitness-based crossover and mutation operators. The second phase is to explore ESN solutions from the surviving individuals by exploiting a sequence of scalarizations using the

polynomial-time algorithm. The obtained nondominated solutions form a front that approximates the true Pareto front.

This study is organized as follows. In Section 2.2, the notations and the bi-objective integer programming formulation of the problem are introduced. We recall the bi-objective branch-and-bound in Section 2.3. The two-phase multi-objective evolutionary algorithm is proposed in Section 2.4. Computational experiments in Section 2.5 evaluate our algorithm by comparing it with three algorithms from the existing literature.

2.2 Problem and definitions

In this section, we introduce a mathematical formulation for the FTL-TPTT and definitions in multi-objective optimization.

The FTL-TPTT considers a procurement auction where m carriers bid for n shipping lanes. Let $I = \{1, 2, \cdots, m\}$ be a set including all the carriers and $J = \{1, 2, \cdots, n\}$ be a set including all the shipping lanes. For each lane $j \in J$, the demanded freight volume, announced by the shipper, is d_j and the capacity of the carrier $i \in I$ is q_{ij}. If a shipping lane $j \in J$ is not operated by a carrier i, $q_{ij} = 0$. Further, the transit time and the bidding rate of the carrier i on lane j are t_{ij} and r_{ij}, respectively. If a carrier i wins a contract, a minimum quantity commitment (MQC) requires that the total freight volume allocated to that carrier should be at least p_i. The goal of the FTL-TPTT is to devise a plan to allocate freight volume on each shipping lane to a set of winning carriers, and minimize both the total transportation costs and the transit time.

Let $M = \max_{i \in I} \sum_{j \in J} q_{ij}$ be the maximum of all carriers' total capacities. Let y_i be a binary variable that equals to 1 if, and only if, carrier $i \in I$ wins a contract from the shipper and x_{ij} be an integer variable denoting the amount of freight volume allocated to carrier $i \in I$ on lane $j \in J$. For convenience, we summarize the notations used below:

Parameters

I the set of carriers, $I = \{1, 2, \cdots, m\}$;

J the set of lanes, $J = \{1, 2, \cdots, n\}$;

d_j the freight volume on lane j;

q_{ij} the capacity of carrier i on lane j;

t_{ij} the transit time of carrier i on lane j;

r_{ij} the transportation rate of carrier i on lane j;

p_i the MQC of carrier i;

Decision variables

x_{ij} an integer variable indicating the freight volume allocated to carrier i on lane j;

y_i a binary variable indicating whether carrier i is a winning carrier.

The formulation of the FTL-TPTT is as follows:

$$\text{(FTL-TPTT)} \quad z_1 = \min \sum_{i \in I} \sum_{j \in J} r_{ij} x_{ij} \tag{2.1}$$

$$z_2 = \min \sum_{i \in I} \sum_{j \in J} t_{ij} x_{ij} \tag{2.2}$$

$$s.t. \quad \sum_{i \in I} x_{ij} \geqslant d_j, \qquad \forall j \in J \tag{2.3}$$

$$p_i y_i \leqslant \sum_{j \in J} x_{ij} \leqslant M y_i, \qquad \forall i \in I \tag{2.4}$$

$$0 \leqslant x_{ij} \leqslant q_{ij} \text{ and } \mathit{Integer}, \quad \forall i \in I, j \in J \tag{2.5}$$

$$y_i \in \{0, 1\}, \qquad \forall i \in I \tag{2.6}$$

The formulation is the same as the one introduced by Hu et al. [40] except that constraints (2.3) are inequalities other than equations. Our formulation allows the purchase of more freight volume than the demand. The equations that force the satisfaction of the demand cause an increase to the infeasible solutions. We thus relax them for more flexibility. Note that if only one objective exists, the problem is the transportation problem with minimum quantity commitments [48].

Although the formulation is not complicated, the problem is

NP-hard [40]. Their bi-objective branch-and-bound method cannot efficiently solve the instances with more than 25 carriers. However, in practice, there could be hundreds of shipping lanes and dozens of carriers in a transportation service procurement auction. Therefore, there is a demand for an effective bi-objective heuristic that can rapidly approximate the Pareto front.

2.3 Bi-objective branch-and-bound

In this section, we recall the bi-objective branch-and-bound algorithm [40], which inspires us in the design of the bi-objective evolutionary algorithm. The bi-objective branch-and-bound algorithm (BBB) is well structured in two levels. Variables y are enumerated in the first level to explore the branch-and-bound tree and variables x are searched, at a leaf node, in the second level to explore ESN solutions.

Let I_0 and I_1 be the index set of carriers with $y_i = 0$ and $y_i = 1$ at a node, respectively. Each node σ in the branch-and-bound tree can be denoted by (I_0^σ, I_1^σ). A leaf node has all the variables y_i fixed to either zero or one. The bi-objective branch-and-bound algorithm implicitly explores all the nodes characterized by variables y in the branch-and-bound. For each leaf node, the nondominated solutions are examined. The framework for the bi-objective branch-and-bound algorithm is shown in Algorithm 2.1.

The algorithm starts from the root node which relaxes all variables y_i to continuous, and then repeatedly computes each node until the node list Λ is empty. For each active node, a binary branching strategy is launched to generate two new child nodes. For each child node, two LP relaxations are solved to generate its local ideal point (z_1^*, z_2^*), where z_1^* and z_2^* are the optimal solution values obtained through computing the linear programming (LP) relaxation problem with a single objective (2.1) and a single objective (2.2), respectively. A node can be pruned if (z_1^*, z_2^*) is dominated by some points in Z_N . If not, it is either branched or solved by CAN method [40] depending on whether it is a leaf node. The obtained feasible integer solutions are used to update the nondominated point set Z_N recursively. Finally, Z_N consists of all the ESN points on the Pareto front.

Algorithm 2.1 Bi-objective branch-and-bound algorithm

$Z_N = \varnothing$, $I_0^{root} = \varnothing$, $I_1^{root} = \varnothing$;
Add root node (I^{root}, I^{root}) into Λ;
while Λ is not empty
    Remove a node σ from Λ;
    Pick a carrier i from $I \setminus (I_0^\sigma \cup I_1^\sigma)$;
    Create child nodes (I_0^σ, $I_1^\sigma \cup \{i\}$) and ($I_0^\sigma \cup \{i\}$, I_1^σ);
    for each child node;
        Solve its LP relaxations to obtain local ideal point (z_1^*, z_2^*);
        if child node is not leaf **and** (z_1^*, z_2^*) is nondominated;
            Add child node to Λ;
        if child node is leaf **and** (z_1^*, z_2^*) is nondominated;
            Solve the subproblem of the child node by CAN;
            Add all the obtained points to Z_N and update;
return Z_N .

2.4 A two-phase evolutionary algorithm

The bi-objective branch-and-bound algorithm enumerates variables *y* in the first phase to explore the branch-and-bound tree and searches variables *x* at leaf nodes in the second phase to explore ESN solutions. Inspired by the two-phase idea, we design a two-phase evolutionary algorithm (TPEA). The difference is that variables y are searched using an evolutionary algorithm in TPEA. Each feasible individual that is characterized by y has a local Pareto front. The local Pareto fronts are further explored in the second phase of TPEA. Instead of exploring all the ESN solutions on the local Pareto fronts of all the individuals, only a few among them with solutions of high quality are explored for efficiency. After the dominance is examined, an approximation front consisting of the obtained nondominated solutions is obtained.

Because of full enumeration, the bi-objective branch-and-bound algorithm is not efficient for large instances, though the final Pareto front can be well exhibited by all the ESN solutions. A drawback is that a large number of nodes in the branch-and-bound tree do not contribute any nondominated solutions on the final Pareto front. On the contrary, they consume a considerable amount of computational time. TPEA attempts to explore only a portion of possible individuals, which correspond to leaf nodes in the branch-and-bound tree. A trade-off is made by TPEA between the quality of the Pareto front and the computational time. From the perspective of practice, such a trade-off is economical as long as the obtained solutions are nondominated or close to nondominated. Because of efficiency with respect to the computational time, large instances can be tackled by TPEA. Algorithm 2.2 describes the framework of TPEA for the FTL-TPTT.

Algorithm 2.2 Two-phase evolutionary algorithm

```
TPEA(N, MaxIter, TimeLimit )
X_E = Ø, popultion = Ø, iter = 0;
Generate N initial solutions and add to popultion;
// Phase 1: explore candidate individuals
while iter ⩽ MaxIter and Time Limit is not reached
    Generate N/2 new individuals using crossover and mutation operators;
    Compute two corner solutions for each individual;
    Add new solutions to popultion;
    Select N solutions among popultion;
    iter = iter + 1;
// Phase 2: explore solutions of individuals
Get a set of individuals 𝒴_E considering the nondominated solutions in popultion;
Explore the nondominated solutions for each individual y ∈ 𝒴_E by CAN;
Add all newly obtained solutions to X_E and update;
return X_E.
```

2.4.1 Individual

While designing an efficient evolutionary algorithm for the FTL-TPTT, it would be difficult to encode a chromosome using all the $m + mn$ decision variables (x, y). Instead, we consider an indirect encoding that only uses m decision variables y to encode a chromosome that specifies an individual. With the encoding scheme, each individual corresponds to a subproblem, where the winning carriers are decided according to the values of y in the chromosome. The subproblem is also bi-objective and has a local Pareto front consisting of solutions that vary the values of x. That is, the chromosome determines a set of winning carriers that will be finally allocated with positive freight volume from the shipper. With the set of winning carriers, the subproblem faced by the shipper is to allocate freight volume to only the winning carriers on all the shipping lanes. As two objectives exist, the subproblem is also bi-objective and can be solved to obtain a local Pareto front.

The problem of an individual can be solved by a dichotomic search method with adapting weights assigned to the two objectives, such as the CAN method presented in Hu et al. [40]. Let λ_1 and λ_2 be the weights assigned to objectives (2.1) and (2.2), respectively. The weighting problem of individual y is formulated as:

$$TP(\lambda_1, \lambda_2, y): \quad \min \sum_{i \in I} \sum_{j \in J} (\lambda_1 r_{ij} + \lambda_2 t_{ij}) x_{ij} \qquad (2.7)$$

$$\text{s.t. constraints} \qquad (2.3)\text{-}(2.6)$$

An individual y is said to be *feasible* if there exists at least one feasible solution x of $TP(\lambda_1, \lambda_2, y)$ for any $\lambda_1 \in \mathbb{R}$ and $\lambda_2 \in \mathbb{R}$. If x is an optimal solution of $TP(\lambda_1, \lambda_2, y)$, then (x, y) must be a feasible and possibly an efficient solution of the FTL-TPTT. Let $\mathscr{Y} = \{0, 1\}^m$ include all possible values to y and $X_E(y)$ denote the set of nondominated solutions of individual y. If individual y is infeasible, then $X_E(y)$ is an empty set. Thus, the Pareto front of the FTL-TPTT can be obtained by computing the local Pareto fronts for $y \in \mathscr{Y}$ and eliminating the dominated solutions.

To enumerate $\mathscr{Y}$ implicitly, using branch-and-bound sharply increases the computational time when the instance size becomes larger. In our method, only individuals that are promising to produce efficient solutions of the FTL-TPTT are explored. As the individuals are only a portion from $\mathscr{Y}$, the computational time reduces significantly. In our case, the problem of distinguishing two individuals is rather important because only better individuals, that could potentially generate efficient solutions of the FTL-TPTT should be explored. We thus define efficiency to evaluate the potentiality of an individual.

2.4.1.1 Efficiency of an individual

The efficiency of an individual indicates whether the individual could generate efficient solutions for the FTL-TPTT. It relies on the chosen set of winning carriers in an individual.

Figure 2.1 shows a local Pareto front of an individual y. Both points A and A' minimize the first objective of the FTL-TPTT that is set with individual y. Among all the solutions that minimize the first objective, the solution that corresponds to point A has the minimum value of the second objective. Similarly, point B in the figure has the minimum value of the first objective among all the solutions that minimize the second objective. The two points *A* and *B* are *corner points* of individual *y*. The two corner points define a rectangle *R*[*A, B*] to contain all the nondominated points on the local Pareto front.

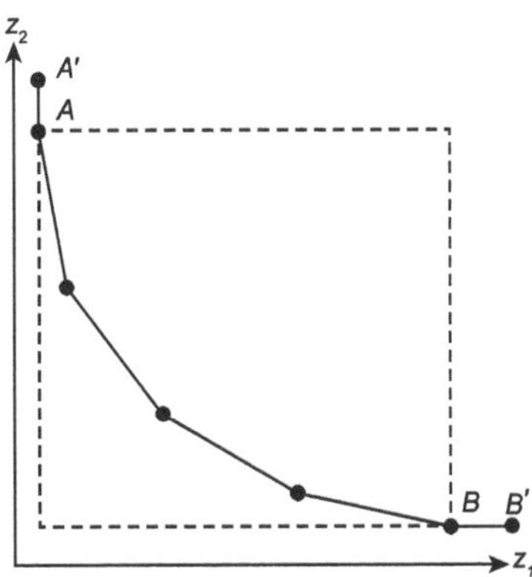

Figure 2.1 Local Pareto front of an individual

The nondominated points on the local Pareto front can be explicitly computed by CAN method [40]. As the possible individuals would be too many, to explore all the local Pareto fronts would be too time-consuming, not to mention that most points from the fronts are just dominated. Therefore, there is no need to compute all the points on a local Pareto front. An imagined curve that connects the two corner points *A* and *B* can be considered to roughly approximate the local Pareto front. An example is given in Figure 2.2. For convenience, if an individual has two corner points *A* and *B*, its imagined curve that represents its local Pareto front is denoted as *A~B*. In Figure 2.2, there are three curves *a~b*, *c~d*, and *e~f* . Dominance among the individuals is clearly shown by the curves. In the figure, the curve *a~b* is nondominated; curve *c~d* are entirely dominated since all points the curve are dominated by some points on curve *a* ~ *b*; curve

$e \sim f$ is partially dominated because part of points on this curve are dominated. A similar technique to imagine the nondominated front has been used in Du and Evans [31].

The imagined curve is realized to be the local Pareto front by exploring the nondominated points on the front. However, there is no need to explore all the nondominated points of an individual at once. In our observation, the number of points could be hundreds or even thousands for some instances. Moreover, all the points on the local Pareto front of many individuals could be just dominated. By using the imagined curves that require only two corner points to represent the local Pareto fronts temporarily, TPEA avoids to explore all the fronts and thus saves computational time. The algorithm only invokes the CAN method to explore nondominated points on the local Pareto fronts for the promising individuals.

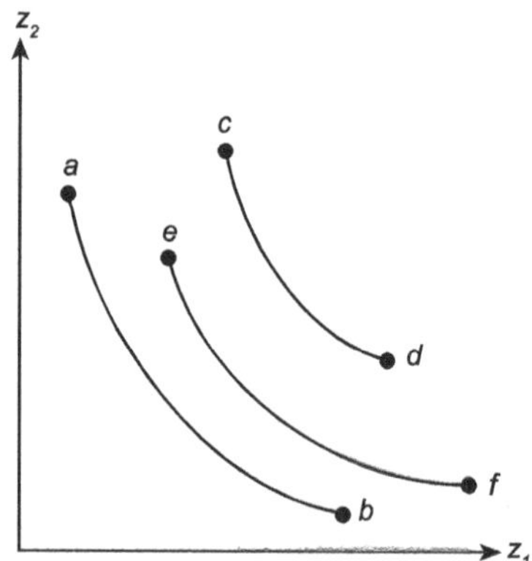

Figure 2.2 Dominance among individuals

An individual is most likely dominated if both corner points are dominated, as observed in the preliminary experiments. Therefore, we simply use the line segment $\overline{AB}$ to represent the imagined curve of an individual in the algorithm, where A and B are the two corner points. Strictly, $\overline{AB}$ is expressed by

$$z_2 = \frac{z_2^B - z_2^A}{z_1^B - z_1^A}(z_1 - z_1^A) + z_2^A, \qquad z_1 \in [z_1^A, z_1^B]$$

where $z_1^A < z_1^B$ and (z_1^A, z_2^A) and (z_1^B, z_2^B) are the coordinates of A and B in the objective space, respectively. In the TPEA, if either corner point is nondominated, the local Pareto front of the individual will be explored.

2.4.1.2 Evaluation of a carrier

As an individual specifies a set of winning carriers, the efficiency of the individual depends on the chosen set of the winning carriers. We evaluate each carrier and design new crossover and mutation operators based on the evaluation scores to generate new individuals.

Generally, few carriers can simultaneously bid low transportation rates and short transit time. Transportation services with shorter transit time are priced higher. In practical applications, a carrier with a larger total capacity has a higher probability of being the winner in the procurement auction, because more capacity implies that the carrier could be more reliable. In addition, the shipper would prefer a carrier with a smaller MQC in order to be more flexible in purchasing desirable transportation services from other carriers.

Therefore, we use MQC and total capacity to evaluate a carrier. Carrier $i \in I$ is evaluated using the following equation:

$$f_c(i) = p_i / \sum_{j \in J} q_{ij} \tag{2.8}$$

A carrier is more preferable to the shipper if the evaluation score is smaller.

2.4.2 Phase one: explore candidate individuals

In the following section, we introduce initialization, crossover, mutation, and selection in the evolutionary algorithm to explore candidate individuals.

2.4.2.1 Initialization

At the beginning of TPEA, a population of solutions is randomly generated. Given a randomly generated individual y, its two corner points are computed by lexicographic optimizations as explained later in Subsection 2.4.3.1. If any of the corner points can be obtained, the individual is feasible and the corner solutions are added into the population. The procedure is repeated until a total of *N* feasible solutions are generated.

2.4.2.2 Crossover

To generate new individuals, a classical multipoint crossover operator (MX) [34] and a fitness-based multipoint crossover operator (FMX) are used. Before either operator is applied, two individuals are obtained, extracted from the two solutions that are randomly chosen from the population. If the two individuals are the same, another two solutions are randomly picked and extracted.

The MX first randomly selects a multiple of crossover points in the sequences of two parent individuals. With the crossover points, the chromosome sequence of each parent individual is divided into several segments. The MX then repeatedly picks a segment from the two parent individuals in turn to construct a new chromosome sequence of a child individual. Another child individual is generated by combining the remaining segments that are not used in the first child individual. In TPEA, the number of crossover points is randomly picked from the interval [1, 6].

The FMX also randomly picks a multiple of crossover points in the sequence. Unlike the MX, which takes segments from the parent individuals in turn, the FMX chooses a segment only from the parent individual that has the better fitness value on the segment. Let S_k be the set of carriers in the *k*-th segment in an individual. The fitness of the segment S_k is defined by the following equation:

$$f_s(S_k) = \sum_{i \in S_k} f_c(i) y_i \Big/ \Big(\sum_{i \in S_k} y_i + \delta\Big) \tag{2.9}$$

here δ is a small number, set to 0.01, to prevent the divisor being zero in TPEA. The fitness of a segment is the average evaluation score over all the winning carriers in the segment. For the *k*-th segment, the FMX chooses the one with the smaller segment fitness value between the two parent individuals. The smaller fitness value indicates larger capacity compared to MQC. Therefore, the newly generated child individual would be feasible with a better chance. Given two parent individuals, the FMX only returns one child individual.

Figure 2.3 illustrates the crossover operation in the FMX. The generated child individual takes the first and the fourth segment from the first parent, and the second and third segment from the second parent. Compared with its parents, the child individual has a relatively small total fitness value.

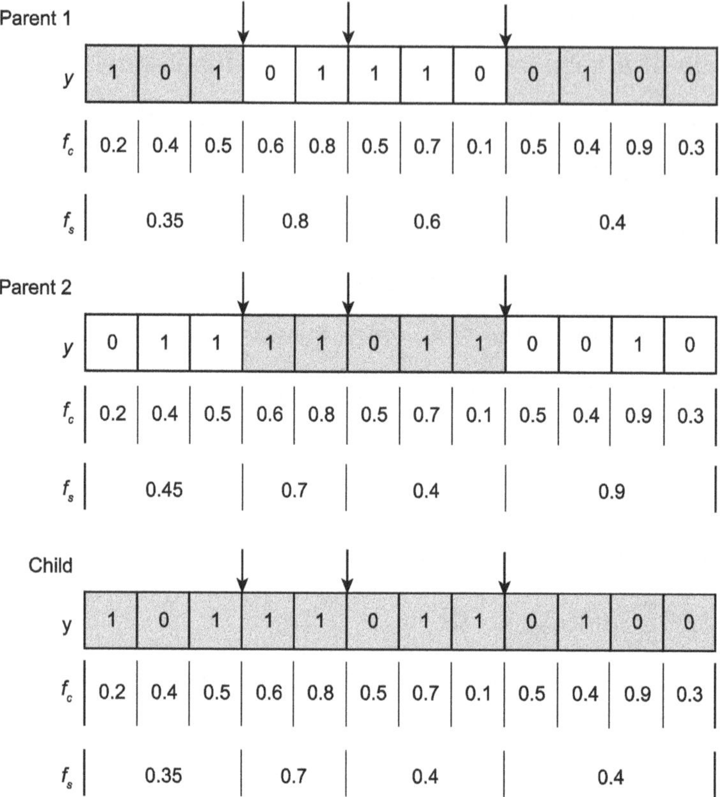

Figure 2.3 An example of the FMX

2.4.2.3 Mutation

Upon mutation, which generates new individuals and extends diversification, a classical bit-flip mutation operator, a fitness-based segment mutation operator, and a fitness-based bit-flip mutation operator are used. Before applying any of the mutation operators an individual is extracted from a solution, that is randomly picked among the population and newly generated solutions.

The classical bit-flip mutation operator serves the purpose of

flipping the value of y_i with a probability for each carrier $i \in I$ in the chromosome of an individual. In TPEA, the probability is set to 0.05.

The fitness-based segment mutation operator serves the purpose of mutating a segment whose fitness is not good enough. Given an individual, the operator will first compute the fitness of each segment in its chromosome. The worst segment that has the largest fitness value is regenerated randomly. Figure 2.4 shows an example of this operator. After selecting the fourth segment with the worst fitness value to make the mutation, the new fitness value decreases.

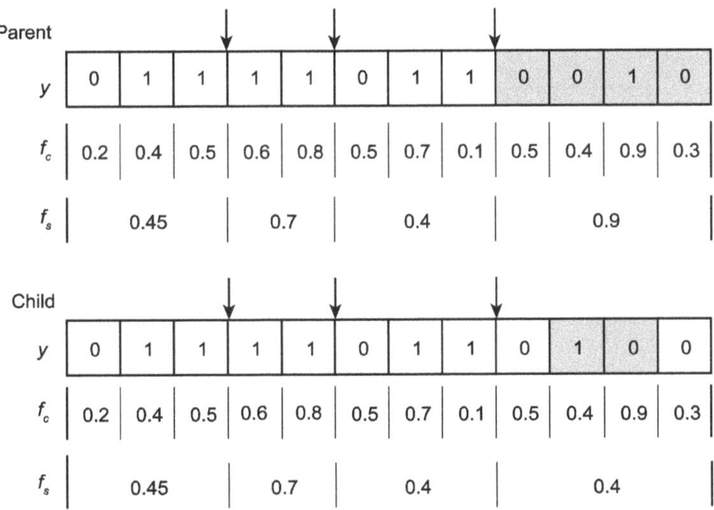

Figure 2.4 An example of the fitness-based segment mutation operator

The fitness-based bit-flip mutation operator adapts the classical bit-flip mutation operator with involving fitness to manage the flip operations. The basic idea is to flip the value of a carrier with good evaluation, to be one with a large probability. The smaller the evaluation score of a carrier, the larger the probability of accepting

the carrier to be a winner. We thus normalize the evaluation scores of all carriers within the interval [0, 1]. The smallest evaluation score among all the carriers is scaled to 1. The normalized evaluation score is assigned to p_i^1, which is a probability that determines whether y_i is set to one. The probability to set y_i to zero is $1-p_i^1$, denoted by p_i^0. A multiple of carriers are chosen at random to set their values to variables y. In detail, if carrier i is chosen for mutation, y_i is set to zero with a probability of p_i^0 if currently y_i = 1 and y_i is set to one with a probability of p_i^1 if currently y_i = 0. An example of this operator is illustrated in Figure 2.5.

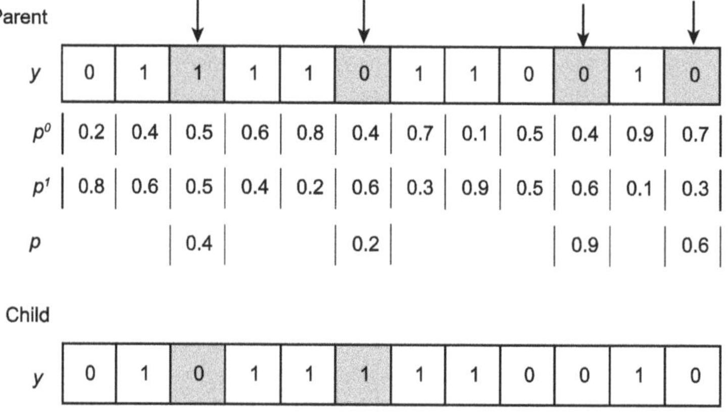

Figure 2.5 An example of the fitness-based bit-flip mutation operator

2.4.2.4 Selection

The selection operator picks N solutions from the population. The intensification and diversification are two significantly important aspects in multi-objective evolutionary algorithms. In TPEA, we implement the classical selection method from NSGA-II[27] that takes both elitism and diversity into account. The selection method in NSGA-II intensifies on elitism by considering the nondomination level of each solution. At the same time, it uses crowding distances for selection so that the retained nondominated points are evenly distributed in the objective space.

2.4.2.5 Enhancements

TPEA continues with a tabu search based on a simple feasibility check and a potentiality check. The infeasible and unpromising individuals are added to a tabu list. If a newly generated individual is already in the tabu list, it will be eliminated immediately.

A feasibility check takes place before solving the two lexicographic optimization problems to obtain two corner points. An individual y is infeasible if there exists a shipping lane $j \in J$ such that the total capacity of all the winning carriers of individual y on the lane does not satisfy the demand, i.e., $\sum_{i \in I} q_{ij} y_i < d_j$. If an individual is infeasible, it will be added into the tabu list and eliminated from the population.

A potentiality check takes place to roughly measure whether an individual can produce efficient solutions of the FTL-TPTT. In our observation, if the total MQCs of all the winning carriers of an individual y is at least twice as large than the total demand of the shipper on all the shipping lanes, there is no need to explore the local Pareto front of the individual. Because the MQCs require the shipper to purchase much more freight volume than the demand, a feasible solution would be dominated, as its objective values could be very large. An unpromising individual is eliminated and forbidden.

2.4.3 Phase two: explore nondominated solutions

In this section, we introduce the procedures that compute the corner points, the nondominated solutions on the local Pareto front of an individual, and the final Pareto front of the FTL-TPTT.

To obtain solutions of an individual y, the integer programming problem $TP(\lambda_1, \lambda_2, y)$ at leaf nodes is frequently solved. However, the problem can be reformulated into a minimum cost flow problem (MCLP), as shown in the following proposition.

Proposition 1. The problem $TP(\lambda_1, \lambda_2, y)$ can be transformed to a minimum cost network flow problem.

Proof. The formulation of the problem $TP(\lambda_1, \lambda_2, y)$ is given in the following:

$$TP(\lambda_1, \lambda_2, y): \quad \min \sum_{i \in I}\sum_{j \in J}(\lambda_1 r_{ij} + \lambda_2 t_{ij})x_{ij}$$

$$\text{s.t.} \sum_{i \in I} x_{ij} \geqslant d_j, \qquad \forall j \in J$$

$$p_i y_i \leqslant \sum_{j \in J} x_{ij} \leqslant M y_i, \qquad \forall i \in I$$

$$0 \leqslant x_{ij} \leqslant q_{ij} \text{ and } Integer, \qquad \forall i \in I, j \in J$$

We consider a minimum cost flow problem (MCLP) defined on a directed graph network $G = (\mathscr{V}, \mathscr{A})$, where $\mathscr{V}$ is a set of nodes and $\mathscr{A}$ is a set of arcs. Each node $i \in \mathscr{V}$ has an external supply b_i. The capacity of each arc $(i, j) \in \mathscr{A}$ is u_{ij}. The cost per unit of flow along arc $(i, j) \in \mathscr{A}$ is c_{ij}. Let f_{ij} be the decision variables that denotes the amount of flow through arc $(i, j) \in \mathscr{A}$. The formulation of the minimum cost flow problem is given as follow:

$$\text{MCLP}: \quad \min \sum_{(i,j) \in \mathscr{A}} c_{ij} f_{ij}$$

$$\text{s.t.} \quad b_i + \sum_{(j,i)\in\mathscr{A}} f_{ji} = \sum_{(i,j)\in\mathscr{A}} f_{ij}, \qquad \forall\, i \in \mathscr{V}$$

$$0 \leqslant f_{ij} \leqslant u_{ij}, \qquad \forall\, (i,j) \in \mathscr{A}$$

The integrality property theorem of the MCLP [1] states that the problem always has an optimal solution that is integer feasible if all the inputs are integer. The network simplex algorithms can compute an integral optimal solution.

According to Figure 2.6, we introduce the network graph of $TP(\lambda_1, \lambda_2, y)$. Let s and w be a source node and a dummy sink node, respectively. Specially, the MCLF that is transformed from $TP(\lambda_1, \lambda_2, \mathrm{y})$ is defined with the following inputs:

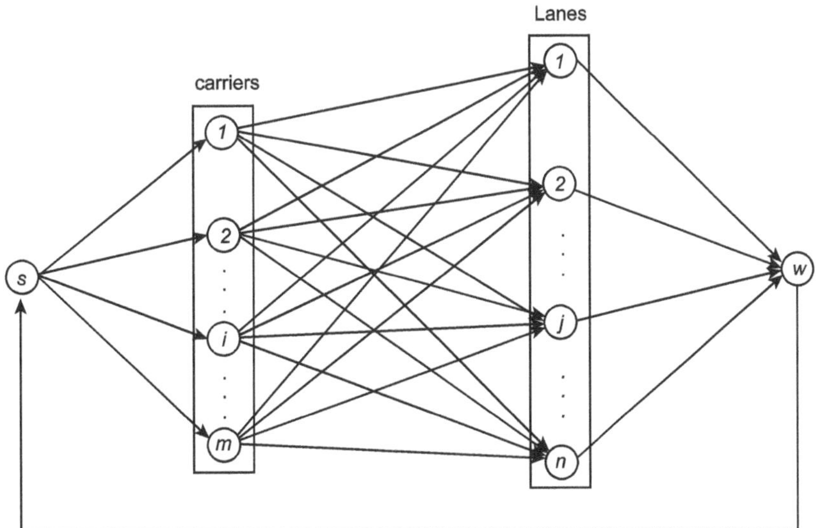

Figure 2.6 An example of a minimum cost network flow model

- $\mathscr{V} = I \cup J \cup \{s, w\}$: the set of nodes;
- $\mathscr{A} = A_1 \cup A_2 \cup A_3 \cup \{(w, s)\}$: the set of arcs, where $A_1 = \{(s, i) : i \in I\}$, $A_2 = \{(i, j) : i \in I, j \in J\}$, and $A_3 = \{(j, w) : j \in J\}$;

- b_v: the external supply on node $v \in \mathscr{V}$;
- u_e: the capacities on arc $e \in \mathscr{V}$;
- c_e: the cost per unit flow through arc $e \in \mathscr{A}$.

$$b_v = \begin{cases} p_v y_v, & \forall_v \in I \\ -d_v, & \forall_v \in J \\ \sum_{j \in J} d_j - \sum_{i \in I} p_i y_i, & v = s \\ 0, & v = w \end{cases}$$

$$u_e = \begin{cases} (M-p_i) y_i, & \forall_e = (s, i) \in A_1 \\ q_{ij}, & \forall_e = (i, j) \in A_2 \\ +\infty, & \forall_e = A \setminus (A_1 \cup A_2) \end{cases}$$

$$c_e = \begin{cases} \lambda_1 r_{ij} + \lambda_2 r_{ij}, & \forall_e = (i, j) \in A_2 \\ 0, & \forall_e \in \mathscr{A} \setminus A_2 \end{cases}$$

Therefore, the problem $TP(\lambda_1, \lambda_2, y)$ can be transformed into the MCLP in polynomial time. Since all the inputs together with λ_1 and λ_2 are integer, the optimal solution of $TP(\lambda_1, \lambda_2, y)$ can be obtained by applying sophisticated polynomial time algorithm to the resulted MCLP. Given an integer optimal solution of the MCLP, an optimal solution of $TP(\lambda_1, \lambda_2, y)$ is obtained, i.e. $x_{ij} = f_{ij}$, $\forall_i \in I, j \in J$.

The integrality theorem of the MCLP [1] states that an optimal solution that are integer always exists as long as all the input data are integers. Network flow algorithms can compute integer optimal solutions for the MCLP. Therefore, the integer program $TP(\lambda_1, \lambda_2, y)$ is efficiently solved. In the implementation, an optimal solution of $TP(\lambda_1, \lambda_2, y)$ is efficiently computed by just applying common solvers such as ILOG CPLEX to solve its linear programming relaxation.

2.4.3.1 Corner points

Given an individual y, its corner points z^1 and z^2 are computed by lexicographic optimization. Let x^* be an optimal solution of the problem *LTP* (1, 0, y). x^* minimizes the first objective of the FTL-TPTT with individual y. However, x^* could be a weak efficient solution of individual y because a multiple of solutions achieving $z_1(x^*)$ could exist. Among all these solutions, only the one that has the minimum value of the second objective defines the first corner point z^1. Therefore, the corner point z^1 is computed by solving another problem which consists of LTP(0, 1, y) and an additional constraint $\sum_{i \in I} \sum_{j \in J} r_{ij} x_{ij} \leqslant z_1(x^*)$. The process of computing the corner point z^1 is lexicographic optimization and is denoted as $z^1 = lexmin_{x \in X(y)}\{z_1(x), z_2(x)\}$, where $X(y)$ is the feasible region of x given by individual *y*. Similarly, the corner point z^2 is computed by lexicographic optimization denoted as $z^2 = lexmin_{x \in X(y)}$ $\{z_2(x), z_1(x)\}$.

2.4.3.2 Nondominated points of an individual

After the two corner points z^1 and z^2 of individual y are computed, CAN algorithm which was first proposed by Aneja and Nair [3] can be applied to explore ESN points on the local Pareto front of the individual. With the two corner points, the third nondominated solution of individual y is obtained by solving $LTP(\lambda_1, \lambda_2, y)$ by adapting $\lambda_1 = z_2^1 - z_2^2$ and $\lambda_2 = z_1^2 - z_1^1$. The weights λ_1 and λ_2 define a new optimization direction in the objective space. Thus, another ESN point z^3 is obtained. If z^3 happens to be z^1 or z^2, there exist no other ESN points of the individual in the curve $z^1 \sim z^2$. Otherwise, the curve $z^1 \sim z^2$ is replaced by two new curves $z^1 \sim z^3$ and $z^3 \sim z^2$. More ESN points can be explored in the new curves by adapting λ_1 and λ_2 accordingly. A detailed implementation of the CAN algorithm is referred to Hu et al. [40]. In TPEA, we do not explore all the nondominated solutions, but explore at most 128 new nondominated solutions for an individual. The obtained points are evenly distributed.

2.4.3.3 Pareto front of the FTL-TPTT

After the individual exploration is terminated in phase one, a procedure of refining the Pareto front of the FTL-TPTT takes place. For all the surviving solutions in the population, their individuals are gathered into a set $\mathscr{Y}_E$. For each individual $y \in \mathscr{Y}_E$, CAN algorithm [40] is used to explore more nondominated solutions on its local Pareto front. The newly obtained solutions are added into X_E. The dominance between any two solutions in X_E is examined and all dominated solutions are removed. Therefore, the solutions remained in X_E form an approximation front of the FTL-TPTT.

2.5 Computational experiment

All algorithms were implemented as a sequential program in Java. The reported computational results were obtained by a PC equipped with an Intel(R) Core(TM) i74790 CPU clocked at 3.60 gigahertz, with 16 gigabytes RAM, and running Windows 8.1 Enterprise. The integer programming solver used was ILOG CPLEX 12.51.

2.5.1 Test instances

Hu et al.[40] introduced a category of 480 small instances. An instance is named "case-*m*-*n*-*q*", where *m* is the number of carriers from the set {10, 15, 20, 25}, *n* is the number of lanes from the set {100, 200, 400, 800}, and *q* indicates the setting on MQCs. The exact algorithm by Hu et al. [40] can explore fast all ESN points for instances that involve no more than 15 carriers, but require thousands of CPU seconds to solve the other test instances. Thus, for the test instances that are easily solved by the exact method, there is no need to apply our evolutionary algorithm. Only the 240 "harder" test instances, which have no less than 20 carriers, are used to test TPEA.

The generator by Hu et al.[40] was used to produce a new category of large instances. For these instances, *m* was chosen from {30, 40, 50, 60, 80} and *n* was chosen from {100, 200, 400, 800}. The minimum quantity committed to carrier *i* was generated from *U* [*D*/

$(0.5m)$, $D/(0.3m)$], where $D = \sum_{j \in J} d_j$. For each pair (m, n), a set of 10 test instances was generated and tagged with "case-*m*-*n*". Thus, 200 new instances were produced.

2.5.2 Performance measures

To evaluate the performance of the bi-objective optimization methods, two wellknown indicators, the hypervolume gap indicator I_H [107] and the multiplicative binary ϵ-indicator I_ϵ [108], are used as measures.

Given an approximation set of points A in the objective space, the hypervolume $H(A)$ represents the volume of the region that is bounded by a reference point. A nadir point is usually the reference point. Considering set B as a reference set, the hypervolume gap indicator $I_H(A, B)$ is computed as:

$$I_H(A, B) = 1 - H(A)/H(B)$$

With the indicator $I_H(A, B)$, smaller values indicate better. The value of this indicator could be negative if A covered more region.

Considering an approximation set A and a reference set B, the multiplicative binary ϵ-indicator $I_\epsilon(A, B)$ is defined as:

$$I_\epsilon(A, B) = \inf_{\epsilon \in \mathbb{R}} \{\forall z^2 \in B, \exists z^1 \in A, z_1^1 \leqslant \epsilon z_1^2, z_2^1 \; \epsilon z_2^2\}$$

It defines the smallest factor ϵ such that for any point z^1 from B, the multiplied point $\epsilon \cdot z^1$ is dominated by at least one point z^2 in A. The smaller the $I_\epsilon(A, B)$, the better the performance of A.

2.5.3 Analysis on operators

We tested two implementations of TPEA to evaluate the crossover and mutation operators methods. The basic implementation (TPEA-1) was the two-phase evolutionary algorithm with only the classical multipoint crossover operator and the classical bit-flip mutation operator. Based on TPEA-1, the second implementation (TPEA-2) was extended to use also the fitness-based crossover and mutation operators. As there would be multiple operators available, each operator was chosen with an equal probability.

A test set was created by including an instance from each data set. The instances that had less than 30 carriers were classified as "small" while the rest were considered to be "large". After adapting the model with constraints (2.3), the bi-objective branchand-bound(BBB) algorithm and the ϵ-constraint method(ϵCM), introduced by Hu et al. [40], were applied with a time limit of 4 hours to solve the small and large instances in the test set, respectively. Their solutions were used as a reference set to evaluate the performance of the implementations of the proposed multi-objective evolutionary algorithm. Each implementation, with a population size N from {50, 100, 200}, was compared with the test set. The time limit for the termination of phase one was not used. The maximum number of iterations was set to $[1/2^{(6-p)} \times B/(2m)]$, where B was the number of bids in an instance and $p \in \{1, 2, 3, 4, 5, 6\}$.

Figure 2.7 shows the performance of each implementation on the test set of small instances using the hypervolume gap indicator for each setting on N . For these small instances, BBB in the literature reported all the ESN points on the Pareto front. The obtained approximation front by each implementation tended to be close to the Pareto front when the number of iterations were increased. By increasing the population size, all the implementations achieved better results for a given p. For both implementations, TPEA-2 performed significantly better than TPEA-1 when both population size and p were small. It implies that the fitness-based crossover and mutation operators are effective in generating new promising individuals. As population size and p increase, both implementations can solve the problem optimally.

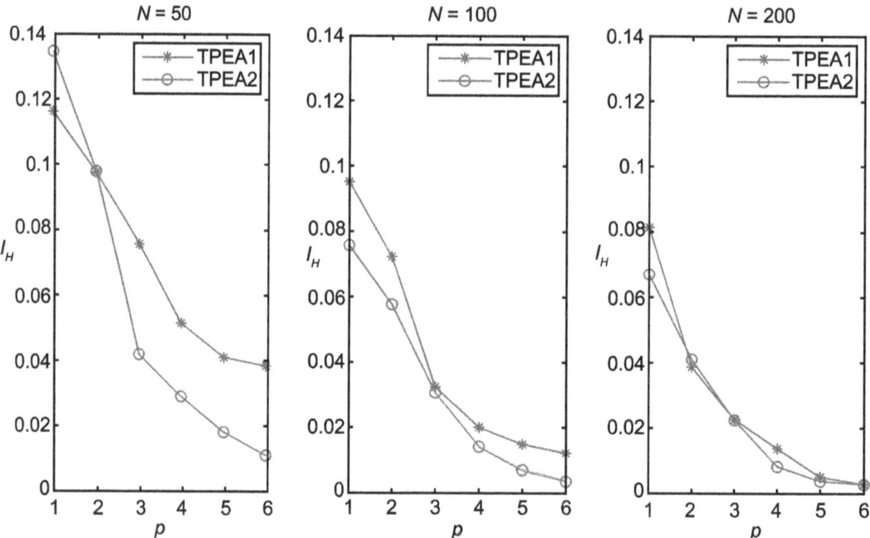

Figure 2.7 Performance on a test set of small FTL-TPTT instances

Figure 2.8 shows the performance of each implementation on the test set of large instances using the ϵ-indicator. The reference set of nondominated points was computed by ϵCM. We did not apply

BBB because it is time-consuming for large instances. Furthermore, it is not sure that the obtained solutions within a time limit are nondominated unless all the nodes in the branch-and-bound tree are implicitly computed. However, each solution computed by ϵCM was indeed nondominated. The shortcoming of ϵCM is that is time-consuming for a large instance to compute a nondominated solution. Although a time limit of 4 hours was set, the obtained nondominated solutions, consisting the reference set, were only a few, ranging from 4 to 10. Therefore, using the hypervolume gap indicator seems unfair for ϵCM. It is possible for a heuristic to compute a number of solutions which could be dominated, as long as they cover more region and contribute a better hypervolume value. Thus, the multiplicative binary ϵ-indicator is used to estimate the distance from the approximation front to the reference set.

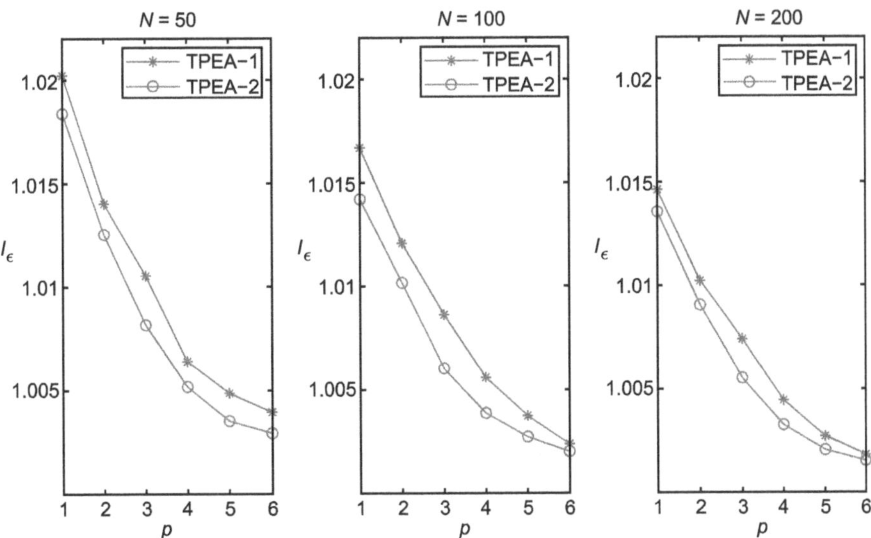

Figure 2.8 Performance on a test set of large FTL-TPTT instances

For all combinations of different population sizes and values of p, TPEA-2 performed much better than TPEA-1, as shown in Figure 2.8. Therefore, we chose to use TPEA-2 with $N = 200$ and $p = 5$ to run all the small instances and TPEA-2 with $N = 100$ and $p = 5$ to run all the large instances considering the trade-off between performance and computational time.

2.5.4 Performance evaluation

The performance of TPEA-2 was first evaluated by comparing its solutions to the nondominated solutions obtained by BBB and ϵCM [40]. Both the hypervolume gap indicator and the multiplicative binary ϵ-indicator were used for evaluation. BBB was applied to solve all the small instances where the number of carriers is less than 30, and ϵCM was applied to all the large instances. For each run, both BBB and ϵCM were set with a time limit of 4 hours. The maximum number of nondominated solutions to be generated by ϵCM was set to 10. Some instances were failed to be solved by BBB or ϵCM within the specific time limit because of complexity.

TPEA-2 was then compared to NSGA-II [27], which is one of the most popular multi-objective heuristics. In the implementation of NSGA-II, it uses the same classical multipoint crossover operator and the bit-flip mutation operator as TPEA-1. However, it differs from TPEA-1 that it has only one phase in the implementation. That is, whenever a feasible individual is found, NSGA-II directly uses CAN algorithm to explore ESN solutions on the local Pareto front. The time limit of NSGA-II was set to the computational time used by TPEA-2 for each instance.

TPEA-2 was tested on any instance that was solved by BBB or ϵCM by obtaining at least one nondominated solution. If an instance had less than 30 carriers, N was set to 200 in TPEA-2, otherwise, N was set to 100. A time limit of 2000 seconds and a maximum of $[B/(4m)]$ iterations were set for the evolution loop in TPEA-2. TPEA-2 was set to exit the loop if no new improved solutions were found within 50 consecutive iterations. Since it takes time for TPEA-2 to explore nondominated solutions, using CAN for surviving individuals in phase two, the computational time would exceed 2000 seconds for some instances.

Table 2.1 reports the comparison results among BBB, NSGA-II and TPEA-2 on small instances. The number of instances solved by BBB in each data set is given in column *#Solved*. Columns #ND and #Sol report the average number of nondominated solutions obtained by BBB and the average of solutions computed by TPEA-2 on the solved instances in a set, respectively. The average computational time in CPU seconds is given in column $\bar{t}(s)$. For TPEA-2, columns $\bar{I}_H$ and $\bar{I}_\epsilon$ indicate the average quality of the obtained approximation front. Table 2.2 reports the comparison results among ϵCM, NSGA-II, and TPEA-2 on large instances using similar columns.

Table 2.1 Comparison results of BBB, NSGA-II and TPEA-2 on small FTL-TPTT instances

Set	#Solved	BBB		NSGA-II (N = 200)				TPEA-2 (N = 200)			
		#ND	$\bar{t}$(s)	#Sol	$\bar{t}$(s)	$\bar{I}_H$	$\bar{I}_\epsilon$	#Sol	$\bar{t}$(s)	$\bar{I}_H$	$\bar{I}_\epsilon$
case-20-100-1	10	126.9	4645.1	113.8	5.6	0.07634	1.00509	211.3	5.5	-0.00054	1.00054
case-20-100-2	10	172.2	7212.5	151.0	8.5	0.10924	1.00750	194.8	8.3	0.00695	1.00151
case-20-100-3	10	161.5	1590.3	126.9	10.3	0.25876	1.02189	160.8	10.1	0.03012	1.00262
case-20-200-1	10	135.7	3146.2	111.3	14.0	0.11335	1.00614	154.9	13.8	-0.00093	1.00033
case-20-200-2	10	219.5	4561.3	156.9	28.3	0.06908	1.00429	212.8	27.8	0.00155	1.00065
case-20-200-3	10	214.8	1154.4	133.7	39.0	0.26405	1.01708	178.9	38.7	0.00439	1.00097
case-20-400-1	10	160.8	2317.4	120.7	37.6	0.09350	1.00558	154.8	37.1	-0.00058	1.00033
case-20-400-2	10	255.3	4252.3	157.3	112.5	0.09369	1.00535	201.6	111.5	0.00077	1.00056
case-20-400-3	10	238.7	1052.2	142.7	118.4	0.31729	1.02909	162.8	117.2	0.00293	1.00068
case-20-800-1	10	163.9	1442.2	123.5	49.9	0.05673	1.00317	126.1	48.9	0.00004	1.00030
case-20-800-2	10	301.4	2526.3	153.8	241.4	0.06611	1.00406	186.3	238.1	0.00091	1.00047
case-20-800-3	10	288.6	551.5	136.9	182.3	0.32912	1.04468	157.3	179.6	0.00653	1.00083
case-25-100-3	9	176.0	8855.3	114.6	17.3	0.45025	1.05324	161.8	17.2	0.03476	1.00231
case-25-200-1	1	116.0	5367.8	112.0	11.0	0.20395	1.01113	180.0	10.9	-0.00229	1.00021
case-25-200-2	1	263.0	13153.0	181.0	21.0	0.14510	1.00842	241.0	20.9	0.00008	1.00045
case-25-200-3	10	229.0	10191.9	138.5	43.4	0.30021	1.02212	175.5	42.9	0.00470	1.00100
case-25-400-1	2	159.5	9933.8	135.5	35.3	0.08511	1.00483	132.0	35.0	-0.00056	1.00028
case-25-400-2	1	305.0	13648.7	136.0	150.4	0.07598	1.00497	306.0	150.3	0.00159	1.00046
case-25-400-3	7	249.6	5216.2	143.3	151.7	0.36233	1.04157	182.0	150.4	0.01038	1.00108
case-25-800-1	4	207.0	5824.9	125.5	143.2	0.05361	1.00301	133.3	143.0	0.00019	1.00034
case-25-800-2	4	331.5	10028.1	172.3	723.1	0.05916	1.00325	196.0	720.5	0.00110	1.00043
case-25-800-3	7	313.0	4928.0	130.3	546.2	0.37137	1.06605	151.9	541.7	0.00049	1.00036
All	166	217.7	5527.2	137.2	122.3	0.17974	1.01693	180.1	121.3	0.00466	1.00076

In Table 2.1, TPEA-2 used much less computational time than BBB and obtained a good approximation front that was very close to the true Pareto front, as indicated by the two indicators. For the sets of some instances, the average hypervolume gap $\bar{I}_H$ is negative because TPEA-2 produced more solutions to cover the region in the objective space. A drawback of BBB is that it takes too much time to examine implicitly all the nodes in the branch-and-bound

tree. However, many nodes in the branch-and-bound tree are not worth being enumerated, as no nondominated solutions could be found. Through creating an individual, which defines a set of winning carriers in TPEA-2, the heuristic searches only some of the leaf nodes in the branch-and-bound tree. By using the fitness-based operators to enumerate new individuals or new leaf nodes, TPEA-2 can approach fast the nondominated solutions on the Pareto front. In addition, TPEA-2 performed better than NSGA-II using almost the same computational time, as shown by $\bar{I}_H$-indicator and $\bar{I}_\epsilon$-indicator. The shortcoming of NSGA-II is similar to that of the BBB, i.e., it might explore the local Pareto front of unpromising individuals who are not likely to generate nondominated solutions. The good quality of the approximation fronts, revealed in Table 2.1, indicates that TPEA-2 can be applied in a practical transportation service procurement system to efficiently explore solutions for the shipper to conduct quick analysis.

For the large instances, where the number of carriers is at least 30 and the number of decision variables could be more than ten thousand, it was very time-consuming for ϵCM to find a nondominated solution. As shown in Table 2.2, ϵCM did not produce any nondominated solution within the time limit for 30 instances, including all the 10 instances from the set “case-80-800”. Furthermore, a few hours were needed to produce only 4 or 10 nondominated solutions for some other instances. However, TPEA-2 still produced hundreds of solutions within acceptable running time. The approximation front can cover more region than ϵCM, as shown by the hypervolume gap indicator. Moreover, the distance from the approximation front to the nondominated points obtained by ϵCM is small, as shown by the ϵ-indicator. Again, TPEA-2 outperformed NSGA-II also for solving large instances.

Table 2.2 Comparison results of ϵ-constraint method, NSGA-II and TPEA-2 on large FTL-TPTT instance

Set	#Solved	ϵCM		NSGA-II (N = 200)				TPEA-2 (N = 200)			
		#ND	$\bar{t}$(s)	#Sol	$\bar{t}$(s)	$\bar{I}_H$	$\bar{I}_\epsilon$	#Sol	$\bar{t}$(s)	$\bar{I}_H$	$\bar{I}_\epsilon$
case-30-100	10	10	50.2	116.6	11.2	0.55247	1.07976	149.3	11.0	-0.03294	1.00295
case-30-200	10	10	168.0	127.5	51.5	0.54743	1.08805	181.6	51.0	-0.05184	1.00136
case-30-400	10	10	1767.5	138.0	314.8	0.38648	1.03894	190.8	313.6	-0.06116	1.00121
case-30-800	10	10	4030.3	145.0	1104.2	0.38404	1.04078	193.5	1097.2	-0.06027	1.00071
case-40-100	10	10	78.9	124.6	20.6	0.48609	1.05616	165.7	20.2	0.00718	1.00541
case-40-200	10	10	468.8	145.2	96.7	0.38066	1.03506	190.4	96.1	-0.03804	1.00189
case-40-400	10	10	1642.5	150.8	499.7	0.28110	1.02514	204.3	496.8	-0.05607	1.00096
case-40-800	6	10	3776.7	139.7	1756.7	0.30092	1.02874	228.3	1751.0	-0.05888	1.00093
case-50-100	10	10	116.0	139.5	27.1	0.45279	1.07691	181.4	26.7	0.02323	1.00542
case-50-200	10	10	487.7	159.9	140.0	0.31259	1.02665	208.5	138.8	-0.03021	1.00275
case-50-400	9	10	4713.7	159.8	690.9	0.28482	1.02431	194.1	688.4	-0.04199	1.00160
case-50-800	6	10	9641.2	160.3	2054.7	0.35236	1.03055	246.2	2042.8	-0.04520	1.00120
case-60-100	10	4	90.0	148.1	44.7	0.19605	1.03146	183.1	44.2	-0.16873	1.00637
case-60-200	10	4	460.0	169.2	198.0	0.16936	1.02419	212.5	196.6	-0.22185	1.00310
case-60-400	8	4	2469.8	193.8	959.6	0.07855	1.01812	224.3	953.8	-0.23080	1.00208
case-60-800	6	4	9883.3	171.5	2071.5	0.13626	1.01985	209.3	2063.4	-0.22245	1.00215
case-80-100	10	4	203.1	169.6	64.2	0.20958	1.02953	195.2	63.5	-0.14983	1.00568
case-80-200	10	4	680.3	185.9	284.2	0.17571	1.02508	209.6	281.9	-0.20140	1.00370
case-80-400	5	4	7584.6	151.0	1610.8	0.10582	1.01855	220.2	1599.8	-0.22369	1.00237
All	170	7.9	2013.8	152.4	631.6	0.30490	1.03778	199.4	628.2	-0.09816	1.00273

2.6 Summary

In this chapter, a multi-objective evolutionary algorithm for the bi-objective full truckload transportation service procurement problem with transit time was studied. Using a set of winning carriers to define individuals makes the algorithm to be efficient, because there is no need to examine all the non-leaf nodes as the bi-objective branch-and-bound algorithm does. The proposed fitness-based crossover and mutation operators are effective to explore new individuals according to the computational results. The experimental comparison results showed that our algorithm is efficient to explore nondominated solutions on the Pareto front. By adapting the models to accommodate with the practical business constraints, such as shipper preferences and carrier number restrictions, the algorithm can be easily extended to solve a class of transportation service procurement problems. The design of the algorithm is also interesting and can be suggested to researchers who also work on the bi-objective problems that have nice structural properties in the subproblems. The algorithm can be integrated into the practical transportation service procurement systems to report fast good solutions and support decision-making for shippers.

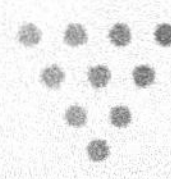

Chapter 3

Bi-objective Branch-and-bound for the Transportation Service Procurement Problem with Transit Time and Total Quantity Discounts

3.1 Introduction

Many companies now outsource transportation and logistics in their value chains to more dedicated players in the transportation industry. Especially, if goods are required to be delivered across the world, they are finally transported by several prominent carriers in the maritime or air transport sector. For some multi-national manufacturers, which are also the global shippers, they operate production bases in the countries with lower labor costs and then supply the markets all over the world. The annual transportation cost of moving raw materials and manufactured products could be multi-billion dollars. With the incentives to reduce costs, centralized procurement activities for purchasing transportation service from carriers are held every year.

In such a procurement activity, a shipper aggregates freight demands and transportation requests from different business units and organizes a multi-round reverse procurement auction. A set of carriers are invited to participate in the auction. At each round, the shipper receives bids from the carriers and runs an optimization engine to generate allocation plans. An allocation plan decides a set of winning carriers and allocates the corresponding freight volume. The optimization engine is run multiple times since different scenarios can be constructed for analysis. Based on the analytical results, the shipper proposes strategies for negotiations or decides the final allocation plan.

In the traditional procurement, the critical objective in the shipper's decision is to minimize transportation cost. However, as revealed by both literature [18] and practice, the non-price factors such as transit time [40], reliability, service quality and carrier reputation [79] also have significant impacts on the shipper's decision. Following Hu et al. [40], we also consider minimizing the transit time in a separate objective in the optimization model. The reasons are two-fold. First, the transit time indicates the service quality level of a carrier from one dimension. By minimizing the transit time, the shipper can seek rapid delivery service for goods. The supply chain would be more agile and meet more expectations from markets. Second, the transit time is a quantitative metric other than a qualitative term. It is suitable to be optimized in a mathematical model. As to other qualitative factors such as service quality and reliability, there is generally no a standard rule to weight a number of key performance indicators in order to quantify them. The choices on which indicators and which weights to use could be different for different procurement managers. Instead of being quantified and formulated in the optimization model, these qualitative factors are more frequently used in evaluating and selecting the carriers for invitations to the auction. With the two aims of minimizing the transportation cost and the transit time simultaneously, we consider solving a bi-objective optimization problem for the shipper.

In the procurement, minimum quantity commitments (MQC) [48] and quantity discounts are effective tactics to reduce cost for the shipper and to improve revenue for the carriers. After a round of the auction, the shipper negotiates with some carriers to request lower transportation rates or discounts. Meanwhile, with the aim of improving revenue, a carrier may request more freight volume to be guaranteed by the shipper in exchange for attractive discounts or lower transportation rates. In the next round of the auction, the carriers update their bids with new bidding rates or discounts. The discounts are called the total quantity discounts (TQD).[37] A total quantity discount is activated only if the total allocated freight volume on all the lanes is equal to or larger than the agreed MQC.

In the multi-round procurement auction, an interesting question is whether to consider only the newly submitted TQDs within the round or to also allow the existing TQDs from previous rounds. The existing TQDs show that the carriers were willing to provide the discounts for the different MQCs and generally may still be willing to accept allocations where the previous TQDs apply. To be more general, we consider a total quantity discount structure that allows a carrier to specify multiple TQDs. Each TQD requires a different MQC and offers a different discount. Hu et al. [40] studied the bi-objective transportation service procurement problem with transit time (TPTT), where an MQC was provided to each carrier but no discount was considered. In their computational experiments, the Pareto front of an instance was analyzed by varying the MQCs and assigning a discount. It was pointed out that the Pareto front under a case was partially dominated by that under another case. If the MQCs and discounts from all the cases are brought in together, the shipper can obtain a better Pareto front, where the nondominated solutions satisfy different MQCs and achieve different discounts from different carriers. Therefore, by introducing a total quantity discount structure for each carrier, it is more flexible for the shipper to make better decisions in the transportation service procurement.

We study the bi-objective transportation service procurement problem with transit time and total quantity discounts (TPTT-TQD). It minimizes the transportation cost and the transit time in two objectives, and decides the winning carriers and freight volume for them in the allocation. Unlike the TPTT, the TPTT-TQD considers a TQD structure for each carrier. The TQD structure defines a number of volume intervals and specifies a discount for each volume interval. If the total freight volume allocated to a carrier enters into a volume interval, the associated discount is activated. For the TPTT-TQD, an implied decision problem is to secure which discount of each carrier by making a selection among the volume intervals. Thus, regarding the complexity, the TPTT-TQD is much more difficult to be solved than the TPTT. Furthermore, unlike that the demand is exactly satisfied in the TPTT, the TPTT-TQD allows to purchase more than demand in order to activate larger discounts. By achieving larger discounts, it is possible to reduce both the transportation cost and the transit time in the objectives. In addition, the surplus freight volume can help the shipper better react to the market changes.

The TPTT-TQD is a bi-objective optimization problem and formulated as a bi-objective integer programming model. We propose a bi-objective branch-and-bound algorithm to solve the problem. To improve the efficiency of the algorithm, lower and upper bound sets is introduced and two stronger fathoming rules are proposed. Two bounding methods with search region reduction are developed to improve the lower bound sets. A hybrid branching method, which firstly branches in the decision space and then branches in the objective space, is applied. To speed up the tree exploration, a hypervolume gap indicator is defined to measure the approximate volume of the search region of each node. Additionally, two acceleration techniques are proposed speedup the algorithm. Based on the generated data simulating the sea freight full container load (FCL) case, computational experiments show that the algorithm is effective in finding nondominated solutions and improving the lower bound sets. Furthermore, it is shown that the decisions can be improved if the discounts and MQCs provided by a carrier in the previous rounds of the auction are included into a TQD structure for next round.

The remaining of the chapter is organized as follows. The formulation of the TPTT-TQD is described in Section 3.2. The bi-objective branch-and-bound algorithm is presented in Section 3.3 and evaluated by computational experiments in Section 3.4. Finally, in Section 3.5, we draw conclusions and suggest directions for future research.

3.2 Problem, formulation and definitions

In the transportation service procurement problem, the shipper first announces its transportation requests and invites the selected carriers to submit bids. Each bid includes its operating lanes, capacities, transit time, and transportation rates. In the TPTT-TQD, carriers can also offer multiple TQDs so that they can assign a different discount for a different MQC. The MQC is the minimum freight quantity to enter an interval specified in the TQD. After collecting all data, the shipper runs an optimization engine to determine the winning carriers and the corresponding freight allocation plan. The shipper expects to secure transportation services of high efficiency and low cost. Thus, the transportation cost and transit time are minimized in two separate objectives.

In the following, we formally introduce the TPTT-TQD with necessary notations and derive an integer programming formulation for it.

In the TPTT-TQD, the shipper has transportation requests on a set of n shipping lanes, denoted by $J = \{1, 2, \cdots, n\}$, and organizes an auction to procure transportation service. A set of m carriers, $I = \{1, 2, \cdots, m\}$, is invited to participate in the auction. The freight volume on lane $j \in J$ demanded by the shipper is d_j. For each carrier $i \in I$, the capacity on lane $j \in J$ is q_{ij}. Particularly, for any lane j that is not operated by carrier i, there is $q_{ij} = 0$. The transportation rate and the transit time of the transportation service provided by carrier $i \in I$ on lane $j \in J$ are r_{ij} and t_{ij}, respectively. In the TQD structure of carrier $i \in I$, a set of discounts, $K = \{1, 2, \cdots, \kappa\}$, is defined. If the total freight volume allocated to carrier i over all the lanes is in the interval $[l_i^k, u_i^k]$, the purchasing cost on the transportation service from carrier i is discounted by α_i^k. Without loss of generality, we assume $u_i^{k-1} \leqslant l_i^k$ and $\alpha_i^{k-1} \leqslant \alpha_i^k$ for $i \in I$ and $k \in K \setminus \{1\}$.

To formulate the problem, we define two sets of decision variables: binary variables y_i^k and integer variables x_{ij}^k. The binary variable y_i^k takes value one if and only if carrier $i \in I$ is a winner in the auction and the total freight volume allocated to the carrier is in interval $k \in K$, and zero otherwise. The variable x_{ij}^k represents the freight volume allocated to the carrier i on lane j when discount k is activated. Moreover, let c_{ij}^k be the discounted transportation rate for carrier $i \in I$ on lane $j \in J$ when discount $k \in K$ is activated, i.e., $c_{ij}^k = (1-\alpha_i^k) r_{ij}$. For convenience, we summarize the notations used below:

Parameters

I the set of carriers, $I = \{1, 2, \cdots, m\}$;

J the set of lanes, $J = \{1, 2, \cdots, n\}$;

K the set of discounts, $K = \{1, 2, \cdots, \kappa\}$;

d_j the freight volume on lane j;

q_{ij} the capacity of carrier i on lane j;

t_{ij} the transit time of carrier i on lane j;

r_{ij} the transportation rate of carrier i on lane j;

α_i^k the discount rate from carrier i when discount k is activated;

l_i^k lower bound of discount interval k from carrier i;

u_i^k upper bound of discount interval k from carrier i;

Decision variables

x_{ij}^k an integer variable indicating the freight volume allocated to carrier i on lane j when discount k is activated;

y_j^k a binary variable indicating carrier i is a winning carrier in the auction and the total freight volume allocated to the carrier is in interval $k \in K$.

Then, the formulation of TPTT-TQD is derived as follows.

$$\text{(TPTT-TQD)}: \quad z_1 = \min \sum_{i \in I} \sum_{j \in J} \sum_{k \in K} c_{ij}^k x_{ij}^k \tag{3.1}$$

$$z_2 = \min \sum_{i \in I} \sum_{j \in J} \sum_{k \in K} t_{ij} x_{ij}^k \tag{3.2}$$

$$\text{s.t.} \sum_{i \in I} \sum_{k \in K} x_{ij}^k \geqslant d_j, \qquad \forall j \in J \tag{3.3}$$

$$\sum_{j \in J} x_{ij}^k - l_i^k y_i^k \geqslant 0, \qquad \forall i \in I, k \in K \tag{3.4}$$

$$\sum_{j \in J} x_{ij}^k - u_i^k y_i^k \leqslant 0, \qquad \forall i \in I, k \in K \tag{3.5}$$

$$\sum_{k \in K} y_i^k \leqslant 1, \qquad \forall i \in I \tag{3.6}$$

$$x_{ij}^k \leqslant q_{ij} y_i^k, \qquad \forall i \in I, j \in J, k \in K \tag{3.7}$$

$$x_{ij}^k \geqslant 0 \text{ and Integer}, \quad \forall i \in I, j \in J, k \in K \tag{3.8}$$

$$y_i^k \in \{0, 1\}, \qquad \forall i \in I, k \in K \tag{3.9}$$

The TPTT-TQD has two minimization objectives (3.1) and (3.2), which minimize the total cost and the total transit time, respectively. Constraints (3.3) guarantee that the transportation demand on each lane is satisfied. Constraints (3.4) and (3.5) ensure that the freight quantity allocated to carrier i is in the interval $[l_i^k, u_i^k]$ if discount k is achieved. Constraints (3.6) require that at most one discount for each carrier can be activated. Constraints (3.7) state that the capacity of each carrier on each lane cannot be violated.

Note that the MQC is defined as the lower endpoint (i.e., l_i^k) of the interval. An MQC is satisfied if and only if the corresponding discount is achieved (Constraints (3.4)), i.e., $y_i^k = 1$. Though the TQD structure may include multiple MQCs and discounts, the MQCs define the intervals and the total allocated freight volume can be in at most one interval (Constraints (3.6)).

3.3 Solution approach

In this section, we introduce a bi-objective branch-and-bound (BBB) algorithm for the TPTT-TQD. The algorithm consists of key components including subproblem at a node (Section 3.3.2), lower bound sets (Section 3.3.3), upper bound sets (Section 3.3.4), fathoming rules (Section 3.3.5), bounding procedures (Section 3.3.6), branching strategy (Section 3.3.7) and acceleration techniques (Section 3.3.8). They are described in the following.

3.3.1 Bi-objective branch-and-bound method

According to Theorem 7 in Goossens et al. [37], for a total quantity discount problem with constraint (3.3), if all variables y_i^k are fixed to either 0 or 1, the LP relaxation problem by relaxing variables x_{ij}^k to continuous can be transformed to a min-cost flow problem and there exists an optimal solution in which all the values of x_{ij}^k values are integers. Therefore, if all binary variables y_i^k have been determined, the TPTT-TQD becomes a bi-objective linear programming (BOLP) problem which can also be solved polynomially, referred to as a *slice problem*. Based on this special structure of TPTT-TQD, we implicitly enumerate the combinations of variables y_i^k in the bi-objective branch-and-bound tree.

Let set $IK = \{(i, k)|i \in I, k \in K\}$ includes all combinations of different carriers and volume intervals. For each branch-and-bound node, the set IK_0 stores all (i, k) pairs with $y_i^k = 0$, and the set IK_1 includes pairs (i, k) with $y_i^k = 1$. The remaining (i, k) pairs associated to undetermined y_i^k are in the set $IK \setminus (IK_0 \cup IK_1)$. The nodes to be explored are temporally put into a node list Λ.

The general procedure of BBB algorithm is illustrated in Algorithm 3.1. The algorithm starts from the root node which relaxes all variables y_i^k to continuous, and then repeatedly computes each node until the node list Λ is empty (Lines 4-13). For a leaf node, in which all variables y_i^k are set to either 0 or 1, the slice problem is actually a BOLP problem which can be easily solved by the CAN method [3, 40]. The obtained feasible integer solutions are used to update the upper bound set U and the nondominated point set Z_N recursively (Lines 6-8). For other active node p, several LP relaxations are solved to generate lower bound sets. Then, the procedure BOUND($\{L_1^p, L_2^p, L_3^p\}$, U) is applied to check whether p can be pruned according to different fathoming rules. If not, a binary branching strategy BRANCH(IK_0^p, IK_1^p, U) is launched to partition the subproblem into two new smaller subproblems, and the corresponding child nodes are added into the node list Λ. Finally, the nondominated points form the Pareto front of TPTT-TQD.

3.3.2 Subproblem

At each node p of the branch-and-bound tree, a relaxed

subproblem L-TPTTTQD(IK_0^p, IK_1^p) is obtained by fixing the corresponding variables y_i^k of pairs (i, k) in IK_0^p and IK_1^p with 0 and 1 respectively whereas relaxing y_i^k to continous for the remaining pairs. By using the weighted sum function with a weight vector $\lambda = (\lambda_1, \lambda_2)$, the relaxed subproblem with a single objective L-TPTT-TQD(IK_0, IK_1, λ) is written as follows.

Algorithm 3.1 Bi-objective Branch-and-Bound (BBB) algorithm

BBB(U)

$IK = \{(i, k) | i \in I, k \in K\}$;

$Z_N = \emptyset$, $IK_0^{root} = \emptyset$, $IK_1^{root} = \emptyset$;

Λ = PUSH(*root*);

while $\Lambda \neq \phi$

  p = SELECT(Λ);

  if $IK \setminus (IK_0^p \cup IK_1^p) == \phi$

    $Z_N(p)$ = CAN(IK_0^p, IK_1^p, U);

    UPDATE(U, Z_N, $Z_N(p)$);

  else

    $\{L_1^p, L_2^p, L_3^p\}$ = L-TPTT-TQD(IK_0^p, IK_1^p);

    prune = BOUND($\{L_1^p, L_2^p, L_3^p\}$, U);

    if *prune* is false

      BRANCH(IK_0^p, IK_1^p, U);

return Z_N.

$$\min \quad \sum_{i \in I} \sum_{j \in J} \sum_{k \in K} (\lambda_1 c_{ij}^k + \lambda_2 t_{ij}) x_{ij}^k \tag{3.10}$$

$$s.t. \ (3.3) - (3.7)$$

$$x_{ij}^{k} \geqslant 0, \qquad \forall\, i \in I, j \in J, k \in K \qquad (3.11)$$

$$y_{i}^{k} = 0, \qquad \forall (i, k) \in IK_0 \qquad (3.12)$$

$$y_{i}^{k} = 1, \qquad \forall (i, k) \in IK_1 \qquad (3.13)$$

$$0 \leqslant y_{i}^{k} \leqslant 1, \qquad \forall (i, k) \in IK \setminus (IK_0 \cup IK_1) \qquad (3.14)$$

This subproblem is used to compute the lower bounds at a non-leaf node and the feasible solutions at a leaf node. By changing different weight vector λ with CAN algorithm, L-TPTT-TQD(IK_0^p, IK_1^p, λ) at node p is iteratively computed to obtain a set of nondominated points, which is denoted by $Z_N(p)$.

3.3.3 Lower bound set

The common way of bound fathoming is to compare a single lower bound (local ideal point) with an upper bound set [60]. Specifically, a node is fathomed if its ideal point is dominated by any point in the upper bound set. The ideal point $\check{z}$ (also denoted by L_1^p in later) is obtained by solving two subproblems with $\lambda^1 = (1, 0)$ and $\lambda^2 = (0, 1)$, namely $\check{z}$ = (L-TPTT-TQD(IK_0^p, IK_1^p, λ^1), L-TPTT-TQD(IK_0^p, IK_1^p, λ^2)).

Ehrgott and Gandibleux [33] introduced the concept of lower bound set in multi-objective combinatorial optimization. To obtain a lower bound set L^p at node p, we need to generate a set of nondominated points $Z_N(p)$ by solving a series of subproblems L-TPTT-TQD(IK_0^p, IK_1^p, λ). The nondominated points set $Z_N(p)$ can be computed by explicit or implicit enumeration. Explicit computation of all nondominated points

$Z_N(p)$ provides the tightest lower bound set. However, it is not efficient in practice due to the large computational burden. We consider to generate an implicit set with a few nondominated points in $Z_N(p)$. Assume all points in $Z_N(p)$ are sorted in ascending order of the first objective value z_1. For two adjacent points z^i and z^{i+1} in $Z_N(p)$, a local ideal point $\check{z}^i$ can be easily obtained. All such local ideal points are included into the second *lower bound set* for node p:

$$L_{2}^{p} = \bigcup_{i=1}^{h-1} \{\check{z}^i = (z_1^i, z_2^{i+1}) | z^i, z^{i+1} \in Z_N(p)\},$$

where h is the cardinality of $Z_N(p)$.

The lower bound set is further improved using segments in Lauth et al. [36], without additional LPs to be computed. If a nondominated point z^0 is obtained by solving subproblem L-TPTT-TQD(IK_0, IK_1, λ) and δ is the objective value, i.e., $\delta = \lambda_1 z^0 + \lambda_2 z_2^0$, we obtain a lower bound ℓ, which is also a segment in the objective space:

$$\ell = \{z \in \mathbb{R}_{\geqslant}^2 \mid \lambda_1 z_1 + \lambda_2 z_2 = \delta\}$$

By changing the weight vector λ, different nondominated points of the subproblem can be obtained and we can compute a set of lower bound segments. Let $L_3^p = \{\ell_1, \ell_2, \cdots, \ell_h\}$ be a *lower bound segment set* of node p. The polyhedron restricted by all the lower bound segments is the feasible space of node p, so called *search region*. Formally, the search region $S(p)$ is given by:

$$S(p) = L_3^p + \mathbb{R}_{\geqslant}^2 = \bigcap_{\ell \in L_3^p} (\ell + \mathbb{R}_{\geqslant}^2).$$

An example of the lower bound set L_3^p is illustrated in Figure 3.1(a). The set $Z_N(p)$ consists of three points: lexicographic minimal points z^1 and z^2, and point z^3. Point z^3 is obtained by solving the relaxed subproblem L-TPTT-TQD(IK_0, IK_1, λ^3) with the weight vector $\lambda^3 = (z_2^1 - z_2^2, z_1^2 - z_1^1)$. We call $\overline{ab}$ an *ideal segment*, where *a* is the intersection point of ℓ_1 and ℓ_3 and *b* is the intersection point of ℓ_3 and ℓ_2. The search region *S*(*p*) is $\overline{ab} + \mathbb{R}^2_{\geqq}$. In this work, we only focus on the single ideal segment case. However, it should be mentioned that the single ideal segment can be extended to multiple ideal segments easily (Figure 3.1(b)) by finding more nondominated points in $Z_N(p)$ such that the lower bound set L_3^p can be enhanced.

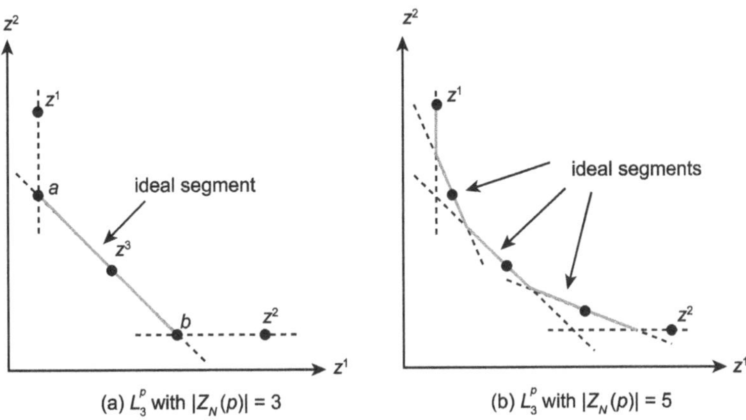

Figure 3.1 Lower bound segment set for a subproblem

3.3.4 Upper bound set

In this section, we describe the method to generate the initial upper bound set, which is further improved with the solutions obtained at the leaf nodes. We also introduce a novel concept to improve the upper bound set.

3.3.4.1 Initial upper bound set

The quality of the upper bound set has a significant impact on the performance of our BBB algorithm. Preliminary computational experiments show that a considerable number of points on the Pareto front actually come from just several leaf nodes of the branch-and-bound tree. The slice problems generating the lexicographic minimal points that are nondominated points may also produce some other nondominated points. Based on this observation, we design a method to generate an initial upper bound set.

The method first computes the lexicographic minimal point z^{TL} of the TPTT-TQD by solving a lexicographic problem, denoted by $lexmin\{z_1(x, y), z_2(x, y)\}$. To solve the lexicographic problem, a commercial integer programming (IP) solver is first applied to minimize the first objective of the TPTT-TQD with obtaining the minimum value z_1^*, then minimize the second objective requiring $z_1 \leqslant z_1^*$. Suppose the solution (x^{TL}, y^{TL}) is a computed solution associated to the lexicographic minimal point z^{TL}. A local Pareto front of feasible points is obtained by invoking the CAN algorithm to compute the slice problem with y^{TL}. Similarly, the lexicographic minimal point z^{BR} is computed by solving lexicographic problem $lexmin\{z_2(x, y), z_1(x, y)\}$, and then another local Pareto front is computed. Finally, an initial upper bound set is constructed by including all feasible points on the two local Pareto fronts.

With the two lexicographic minimal points and the other

nondominated points in the upper bound set, several fathoming rules (Section 3.3.5) are carried out for an attempt to prune the node in the branch-and-bound tree.

3.3.4.2 Improved upper bound set

Let Z_N be a set of nondominated points found so far, including the lexicographic minimal points z^{TL} and z^{BR}. The points in Z_N are ordered such that $z_1^i \leqslant z_1^{i+1}$ for all $i \in \{1, 2, \cdots, |Z_N| - 1\}$. Given two points z^i and z^j, the point $(\max\{z_1^i, z_1^j\}, \max\{z_2^i, z_2^j\})$ is a *local nadir point*. The first upper bound set, denoted by U_N, includes all the local nadir points for any two consecutive nondominated points in Z_N, i.e.,

$$U_N = \{(z_1^{i+1}, z_2^i) \in \mathbb{R}^2 | z^i, z^{i+1} \in Z_N, i \in \{1, 2, \cdots, |Z_N| - 1\}\}$$

The nondominated points are obtained from slice problems, which are BOLP problems. The local Pareto front of such a slice problem consists of ESN points and NSN points. Particularly, the NSN points are on the segment connecting two ESN points. For any two consecutive nondominated points $z^i, z^{i+1} \in Z_N$, we call the segment $\overline{z^i z^{i+1}}$ connecting the two points a *leaf edge*, if the two consecutive points are obtained from the same slice problem. Note that leaf edges are also lower bound segments at leaf nodes. The upper bound set can be improved by replacing some nadir points with the corresponding leaf edges. Let E be the set including all the leaf edges obtained from Z_N and V be set of the replaced nadir points. Thus, the second upper bound set, denoted by U_N^+, consisting of nadir points and leaf edges, i.e., $U_N^+ = (U_N \setminus V) \cup E$.

In Figure 3.2, there are two local Pareto fronts P (a) and P (b)

generated from two slice problems, which contain nondominated points in sets $\{a^1, a^2, a^3, a^4, a^5\}$ and $\{b^1, b^2, b^3, b^4, b^5\}$ respectively. Points a^1 and b^5 are lexicographic minimal points z^{TL} and z^{BR} respectively.

Since point b^1 is dominated by point a^5, Z_N includes all the points on the local Pareto fronts except b^1, i.e., $Z_N = \{a^1, a^2, a^3, a^4, a^5, b^2, b^3, b^4, b^5\}$. For the two consecutive points a^5 and b^2 in Z_N, their nadir point c is considered because $\overline{a^5b^2}$ is not a valid leaf edge. Therefore, we have $U_N^+ = \{\overline{a^1a^2}, \overline{a^2a^3}, \overline{a^3a^4}, \overline{a^4a^5}, c, \overline{b^2b^3}, \overline{b^3b^4}, \overline{b^4b^5}\}$.

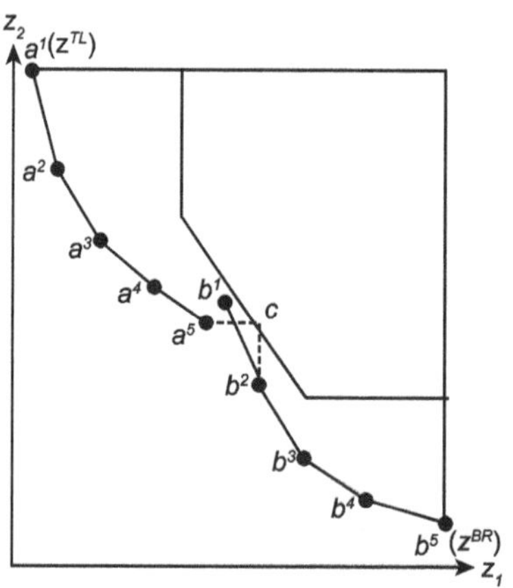

Figure 3.2 An example of upper bound sets

Furthermore, instead of using any point, we consider to define an upper bound set using only the nondominated leaf edges. A leaf edge is *nondominated* if there exist any nondominated points on it. Let U_E denote the upper bound set including all the leaf edges which are nondominated.

Considering the example shown in Figure 3.2, U_E includes leaf edge $\overline{b^1b^2}$ instead of nadir point c, i.e., $U_E = \{\overline{a^1a^2}, \overline{a^2a^3}, \overline{a^3a^4}, \overline{a^4a^5}, \overline{b^1b^2}, \overline{b^2b^3}, \overline{b^3b^4}, \overline{b^4b^5}\}$.

For a slice problem, the CAN method is used to explore a set of leaf edges. For each two consecutive points in the local Pareto front, a leaf edge connecting the two points is constructed and include it into U_E if nondominated. The CAN method iteratively split a leaf edge into two new leaf edges by finding a new nondominated point until no more leaf edge can be split. Each edge can be split at most once. For a leaf edge that cannot be split, it is indivisible and the points on it are supported nondominated points for the slice problem.

3.3.5 Fathoming rules

In this section, we introduce five fathoming rules listed from weak to strong. The Fathoming Rule (1) was proposed by Mavrotas et al.[60] and the Fathoming Rules (2) and (3) were proposed by Gadegaard et al.[36]. In addition, we design two stronger Fathoming Rules (4) and(5) which are based on the improved lower bound sets and upper bound sets.

The basic Fathoming Rule (1) uses only an ideal point as the lower bound L_1^p. By using the lower bound set, we obtain the Fathoming Rule (2).

Fathoming Rule (1). Given the set of nondominated points Z_N, node p, and its ideal point $\check{z}$, node p can be fathomed if $\check{z}$ is dominated by some point $z \in Z_N$.

Fathoming Rule (2) Given the set of nondominated points Z_N, node p, and its lower bound set L_2^p, node p can be fathomed if each local ideal point $\check{z}^i \in L_2^p$ is dominated by some point $z \in Z_N$.

The Fathoming Rule (3) is based on the upper bound set U_N and lower bound segment set L_3^p. It indicates that a node p can be fathomed if all the local nadir points in U_N are outside the search region $S(p)$. In other words, the node p cannot be fathomed if there exists a nadir point of U_N within the search region $S(p)$. Furthermore, we propose stronger Fathoming Rules (4) and (5) using the upper bound set U_N^+ and the leaf edge set U_E, respectively.

Fathoming Rule (3) Given the upper bound set U_N , node p, and its lower bound segment set $L_3^p = \{\ell_1, \ell_2, \cdots, \ell_h\}$, node p can be fathomed if $S(p) \cap U_N = \emptyset$.

Fathoming Rule (4) Given the upper bound set U_N^+, node p and its lower bound segment set $L_3^p = \{\ell_1, \ell_2, \cdots, \ell_h\}$, node p can be fathomed if $S(p) \cap U_N^+ = \emptyset$.

Fathoming Rule (5) Given an upper bound set U_E, node p, and its lower bound segment set $L_3^p = \{\ell_1, \ell_2, \cdots, \ell_h\}$, node p can be fathomed if $S(p) \subseteq \bigcup_{e \in U_E} (e + \mathbb{R}^2_{\geqq})$.

It shows that if all local nadir points and leaf edges in U_N^+ are outside the search region $S(p)$, node p can be fathomed. If the search region $S(p)$ is covered by the subregions dominated by the leaf edges in U_E, node p can be fathomed. The example in Figure 3.2 shows that the Fathoming Rule (5) is stronger than (4). Although all the edges in U_N^+ are outside the search region $S(p)$, the nadir point c is in the search region $S(p)$ and hence we have $S(p) \cap U_N^+ \neq \emptyset$. Instead, U_E includes leaf edge $\overline{b^1 b^2}$ rather than the nadir point c. We obtain $S(p) \subseteq \bigcup_{e \in U_E} (e + \mathbb{R}^2_{\geqq})$. Therefore, node p can be fathomed using Fathoming Rule (5) instead of fathoming rule (4).

In the BBB algorithm, the Fathoming Rules (1)–(5) are used in turn in the bounding procedure. If a node is fathomed by a specific fathoming rule, there is no need to compute the improved lower and upper bound sets and apply the stronger fathoming rules.

3.3.6 Bounding strategies

In this section, we develop two bounding methods implementing the Fathoming Rules (4) and (5). For a node that is not pruned, the bounding methods can reduce the search region by iteratively improving the upper bounds on the two objectives. The reduction can be propagated to its child nodes. Since there are no computational efforts on exploring points in the dominated regions after reduction, the performance of the algorithm is improved.

3.3.6.1 Bounding with search region reduction (Bounding-SRR)

Considering that the lower bound set L_3^p consists of segments rather than nadir points, it is challenging to check the sufficient conditions in Fathoming Rules (3) and (4). Gadegaard et al. [36] implemented Fathoming Rule (3) by checking if any point of U_N is in the search region $S(p)$. To do so, they solved a linear formulation for each point of U_N. This method has two main disadvantages. First, a number of linear programs have to be solved. Second, for each child of node p, solutions outside the search region of node p may be explored.

To overcome these weaknesses, we propose a bounding method which implements Fathoming Rule (4) and reduces search region as well, named as bounding with search region reduction (Bounding-SRR). The basic idea is motivated by the observation that a large part of the search region $S(p)$ is dominated by the upper bound set. For each node, if there is a nadir point or leaf edge in U_N^+ outside the search region $S(p)$, the upper bounds of the objectives might be reduced. If so, the search region $S(p)$ is reduced and propagated to its child nodes. The upper bounds on the objectives are iteratively tightened and the search region $S(p)$ becomes smaller gradually. Finally, if the search region $S(p)$ is empty, the node p is fathomed.

After subtracting the dominated part from the search region, if remaining search region $S(p) - S$ is not empty, it can be continuous or noncontinuous. Our Bounding SRR method focuses on dealing with the continuous search region, whereas the noncontinuous case is taken into account in the branching part introduced later.

In order to implement the Bounding-SRR method, we define a new model TPTTTQD($IK_{0,}$ $IK_{1,}$ ub_1, ub_2) at each node which is a modification of TPTT-TQD($IK_{0,}$ IK_1) by adding two new objective constraints (3.15) and (3.16).

s.t. (3.3) −(3.7), (3.11) −(3.14)

$$z_1 \leqslant ub_1 \tag{3.15}$$

$$z_2 \leqslant ub_2 \tag{3.16}$$

In particular, ub_1 and ub_2 at the root node are the z_1-value of lexicographic minimal point z^{BR} and z_2-value of lexicographic minimal point z^{TL}, respectively. For any other node, ub_1 and ub_2 are inherited from its parent node, and tightened gradually with Algorithm 3.2.

The details of our Bounding-SRR algorithm (Algorithm 3.2) are presented as follows. We first check whether node p is dominated and at the same time update the upper bound of the second objective ub_2' if possible (Lines 2—11). Specifically, each element (point or edge) in U_N^+ is checked in sequence whether it is outside the search region S(p). It checks for each segment in L_3^p whether the point or edge is outside the space dominated by the segment (Lines 4—6). If so, a part of search region is discarded by updating ub_2' (Line 11). Otherwise, the node cannot be pruned and the updating process of ub_2' is terminated (Lines 7—9). Note that if all the nadir points and leaf edges in U_N^+ are outside the search region S(p), the node is fathomed. If the node is not fathomed, we update ub_1' in a similar way except that the elements of U_N^+ are checked in the reverse order.

Algorithm 3.2 Bounding-SRR

BOUNDING-SRR(p, L_3^p, U_N^+, ub_1', ub_2')

1 dom = true, $ub_1' = ub_1$, $ub_2' = ub_2$;

2 **for** element $t \in U_N^+$

3   *outside* = false;

4   **for** segment $\ell \in L_3^p$

5   **if** element t is outside the region $\ell + \mathbb{R}^2_{\geqq}$

6   *outside* = true;

7   **if** *outside is* false

8   *dom* = false;

9   **break**;

Continued

```
        else
                update ub_2' with element t;
if dom = true;
        return;
for element t ∈ REVERSE(U_N^+)
        outside = false;
        for segment ℓ ∈ L_3^p
                if element t is outside the region ℓ + ℝ^2_≧
                        outside = true;
        if outside is false
                break;
        else
                update ub_1' with element t;
return dom, ub_1', ub_2'.
```

The Bounding-SRR method is illustrated in Figure 3.3. At the beginning, the search region $S(p)$ of node p (shaded in dark gray) is a polygon constrained by all the lower bound segments in L_3^p and constraints $z_1 \leqslant ub_1$ and $z_2 \leqslant ub_2$. For the example shown in the figure, node p cannot be fathomed because several local nadir points in U_N^+ lie in the search region S(p). However, the bounds ub_1 and ub_2 are iteratively updated and finally reduced to ub_1' and ub_2', respectively. The search region $S(p)$ is reduced to a smaller one that is constrained by $z_1 \leqslant ub_1'$ and $z_2 \leqslant ub_2'$ and shaded in light gray in the figure.

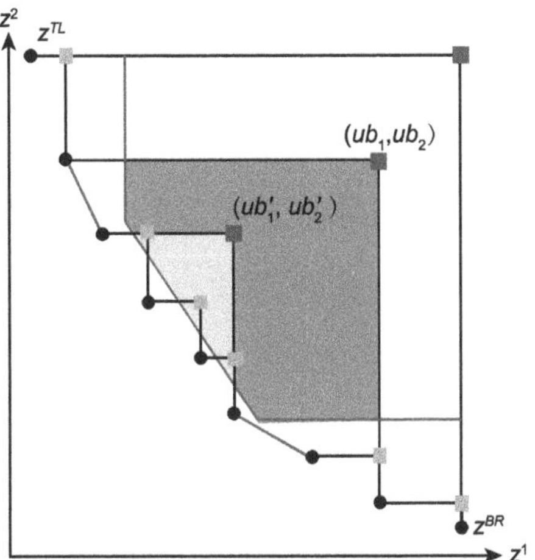

Figure 3.3 Bounding-SRR method

3.3.6.2 Bounding-FS

We also introduce a filter segment bounding (Bounding-FS) method to implement Fathoming Rule (5). The implementation is more challenging compared to that of Fathoming Rule (4), because both the lower bound set L_3^p and the upper bound set U_E include segments instead of points.

Considering the lower bound set shown in Figure 3.1(a), the search region of node p is equivalent to the space dominated by the ideal segment $\overline{ab}$. The node is fathomed if and only if $\overline{ab}$ is dominated. If the ideal segment $\overline{ab}$ is decomposed into a set of small subsegments, the node is fathomed if each subsegment is dominated [74]. Instead of filtering a segment with the points in set Z_N [74], leaf edges in set U_E are used in our Bounding-FS method which is shown in Algorithm 3.3.

Assume points a^e and b^e are the two endpoints of a leaf edge $e \in U_E$, if point a is dominated by the leaf edge e, some part of the lower bound segment $\overline{ab}$ is filtered. Then two cases are considered: (1) if e is outside $\overline{ab} + \mathbb{R}^2_{\geqq}$, ub'_2 is updated to the z_2-value of point b^e; and (2) if e intersects with $\overline{ab}$, ub'_2 is updated to the z_2-value of the intersection point.

With this new ub'_2 value, we can update point a to filter part of segment $\overline{ab}$. The procedure is repeatedly executed until ub'_2 cannot be reduced any more or segment $\overline{ab}$ is filtered to empty so that the node can be fathomed. If $\overline{ab}$ is not empty, we then iteratively improve ub'_1 in a similar way (Lines 4-24).

Algorithm 3.3 Bounding-FS

```
BOUNDING-FS(p, ab, U_E, ub_1, ub_2)
1     dom = true;
2     ub'_1 = ub_1;
3     ub'_2 = ub_2;
4     while ab is not empty
5          canUpdate = false;
6          for leaf edge e ∈ U_E
7               if e ≺ a
8                    canUpdate = true;
9                    update ub'_2 and segment ab;
10                   break;
11         if canUpdate is false
12              dom = false;
13         break;
```

Continued

if *dom* is true
    return;
while $\overline{ab}$ is not empty
    canUpdate = false;
    for leaf edge $e \in U_E$
        if $e \prec b$
            canUpdate = true;
            update ub_1' and segment $\overline{ab}$;
            break;
    if *canUpdate* is false
        break;
return *dom*, ub_1', ub_2'.

The Bounding-FS algorithm can be used alone or after the Bounding-SRR algorithm. In addition, the Bounding-FS algorithm can be easily extended to the case where multiple ideal segments exist.

3.3.7 Branching strategy

We develop a new hybrid branching (HB) strategy that first uses the variable branching to branch in the decision space and then uses the Pareto branching [88] to branch in the objective space. Considering node p and its ideal segment $\overline{ab}$, a set of consecutive nondominated points, denoted by $\overline{Z}_N(p) = \{z^i, z^{i+1}, \cdots, z^{i+j}\}$, can be obtained from Z_N such that the segment $\overline{a'b'}$ is dominated, where points a' and b' are on the segment $\overline{ab}$ with $z_1^{a'} = z_1^i$ and $z_2^{b'} = z_2^{i+j}$. When the dominance is examined, the improved upper bound set U_N^+ is considered. In the Pareto branching, node p is split into two child nodes: one node imposes a new upper bound that is z_1^i on z_1 and a new upper bound that is z_2^{i+j} on z_2 is used for the other node. There is no need to proceed the Pareto branching whenever it is possible. We introduce a distance ratio σ and use the parameter to determine whether to proceed the Pareto branching or not. We set σ to 0.3 in this study.

The hybrid branching strategy is outlined in Algorithm 3.4. In the algorithm, D_1 and D_2 are the distances between lexicographic minimal points z^{TL} and z^{BR} along the two objectives, respectively. The algorithm first applies a standard variable branching by choosing a pair (i, k) from set $IK/(IK_0 \cup IK_1)$ whose decimal portion of the value of variable y_i^k is the closest to 0.5. The incumbent subproblem is then divided into two new subproblems by adding the pair (i, k) into set IK_0 and IK_1, respectively (Lines 2—3). For each child node, we compute the set $\overline{Z}_N(p)$ by checking whether there exists a sequence of points in Z_N whose nadir points and possible leaf edges are outside the search region $S(p)$. The distances between the first and the last point in $\overline{Z}_N(p)$ on the two objectives are denoted as d_1 and d_2, respectively. If any distance ratios, i.e., $\frac{d_1}{D_1}$ and $\frac{d_2}{D_2}$, are larger than σ, the node is further branched with the Pareto branching.

In the tree exploration, a best-first strategy (BFS) is used to select a node to branch in the bi-objective branch and bound method. In the branching with BFS, the node with the largest search region is selected and branched.

Algorithm 3.4 Branching strategy

BRANCH(IK_0, IK_1, ub_1, ub_2, D_1, D_2, σ)

// branch on decision space;

pick a pair(i, k) from $(IK \setminus (IK_0 \cup IK_1))$;

children $\leftarrow$ TPTT-TQD($IK_0 \cup \{(i, k)\}$, IK_1, ub_1, ub_2);

children $\leftarrow$ TPTT-TQD(IK_0, $IK_1 \cup \{(i, k)\}$, ub_1, ub_2);

for node p = TPTT-TQD(IK_0', IK_1', ub_1, ub_2) $\in$ *children*

choose $\overline{Z}_N(p) = \{z^i, z^{i+1}, \cdots, z^{i+j}\}$ accordingly;

$d_1 = \overline{z}_1^{i+j} - \overline{z}_1^i$; $d_2 = \overline{z}_2^i - \overline{z}_2^{i+j}$;

if $\frac{d_1}{D_1} \geqslant \sigma$ **or** $\frac{d_2}{D_2} \geqslant \sigma$

// branch on objective space;

children $\leftarrow$ TPTT-TQD(IK_0', IK_1', z_1^i, ub_2);

children $\leftarrow$ TPTT-TQD(IK_0', IK_1', ub_1, z_2^{i+j});

remove p from children;

return children.

To evaluate the search region for node p, we need to measure the dominated areas for both the upper bound set U and the lower bound set L_3^p. First, the well-known hypervolume indicator [107], denoted by $H(U)$, is used to measure the volume of the region which is dominated by the given upper bound set U and bounded by a reference point r. The reference point r we choose is the nadir point of lexicographic minimal points z^{TL} and z^{BR}.

Since the lower bound segment set L_3^p consists of segments rather than points, the classical hypervolume is no longer suitable.

Therefore, we define an adapted hypervolume indicator $\overline{H}(L_3^p)$ to measure the volume of the region dominated by the segments of L_3^p and bounded by the reference point r, which is computed as follows:

$$\overline{H}(L_3^p) = (z_1^r - z_1^a)(z_2^r - z_2^b) - \frac{1}{2}(z_1^b - z_1^a)(z_2^a - z_2^b)$$

The difference of $\overline{H}(L_3^p)$ and $H(U)$ provides an approximated volume of the search region $S(p)$, i.e.,

$$H(p) = \overline{H}(L_3^p) - H(U)$$

Note that the search region $S(p)$ is not precisely measured, because some nondominated points are outside the the search region. Figure 3.4 shows an approximation of the search region. We find that a more accurate approximation of the search region $S(p)$ can be obtained if we refine the reference point r to point (ub_1, ub_2) in the figure, where ub_1 and ub_2 are computed in the bounding (Section 3.3.6). In this way, $H(U)$ (in gray with line) is reduced by discarding some areas outside the lower bound segments of L_3^p.

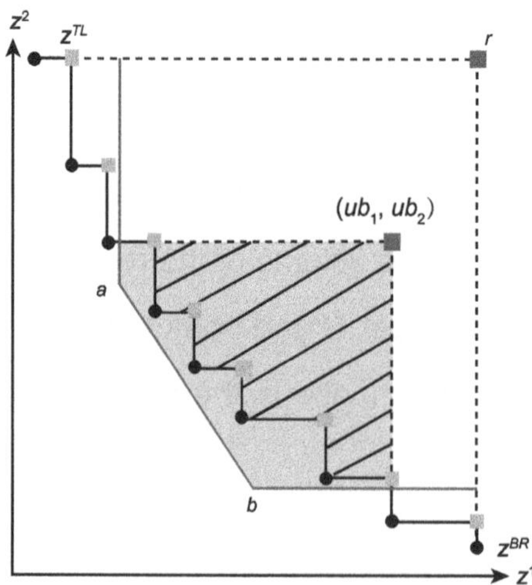

Figure 3.4 Hypervolume gap indicator

3.3.8 Acceleration techniques

To speed up the proposed algorithm, we implemented the following two techniques.

3.3.8.1 Preprocessing on nodes

A simple but effective check could be used to examine the demand feasibility (Constraints (3.3)) at the beginning of each node in the search tree. For a given lane j, the procured quantity from carrier i cannot exceed not only the capacity on lane j, but also the upper bound of the highest volume interval k where $y_i^k \neq 0$ for all $k \in K$. In particular, we cannot procure any freight from carrier i if $y_i^k = 0$ for all $k \in K$. Therefore, we can simply check whether the available freight volume is sufficient to satisfy the demand of the shipper on each lane:

$$\sum_{i \in I} \min\{q_{ij}, u_i^{K(i)}\} \geqslant d_j, \quad \forall j \in J \tag{3.17}$$

where $K(i) = \max\{k | y_i^k \neq 0\}$.

3.3.8.2 Computing lower bound set at non-leaf nodes

Recall that we have to solve four relaxed single-objective LP subproblems to find lexicographic minimal points z^{TL} and z^{BR} of the subproblem L-TPTT-TQD(IK_0^p, IK_1^p) at node p.

The purposes of solving lexicographic problems at leaf and non-leaf nodes are different. We must obtain efficient solutions at leaf nodes, thus it is necessary to solve lexicographic problems. But we only aim to compute a lower bound segment set at non-leaf nodes.

By minimizing the two objectives separately at a non-leaf node, we solve two LPs to obtain two points z^1 and z^2, which are no need to be lexicographic minimal points. The third point z^3 is obtained by solving the relaxed subproblem L-TPTT-TQD(IK_0^p, IK_1^p, λ^3) with the weight vector $\lambda^3 = (z_2^1 - z_2^2, z_1^2 - z_1^1)$. With the minimial objective values z_1^1 and z_2^2, the point z^3, and the weight vector λ^3, a lower bound segment set for the non-leaf node is determined. Therefore, we only solve three LPs instead of five LPs at each non-leaf node and the total computational time is reduced greatly. Preliminary experiments show that this strategy can save up to 50% of the computational time.

3.4 Computational experiments

All algorithms were implemented as a sequential program in Java. The computational results reported were obtained on a PC equipped with an Intel(R) Core(TM) i7-6700 CPU clocked at 3.41 gigahertz, with 32 gigabytes RAM, and running Windows 10 Education. The integer programming solver used was ILOG CPLEX 12.51. The time limit was set to three hours for each instance.

3.4.1 Test instances

We generated a set of test instances to simulate the transportation service procurement problem in the case of sea freight FCL. In the data generation, the number of carriers m was chosen from {10, 15, 20}, and the number of lanes n was chosen from {100, 200, 400}. The number of volume intervals in the TQD structure κ was set to 3 for all carriers. Thus, the number $|IK|$ ranges from 30 to 60. The complexity of the bi-objective branch-and-bound tree is much larger than that of Hu et al.[40].

To generate an instance, for each carrier $i \in I$, the number of its operating lanes was randomly chosen from $[0.3n, 0.7n]$. The transit times t_{ij} on lane j were randomly generated from the interval $[\bar{t}_j (1 - \delta_j), \bar{t}_j (1 + \delta_j)]$, where $\bar{t}_j$ was a mean obtained from [10, 51] and δ_j was the deviation generated from [0.05, 0.15]. Similarly, the transportation rates on each lane were generated with a mean from [20, 101] and a deviation from [0.05, 0.30]. The freight demand on lane j was generated as $d_j \in [10, 200]$, and the capacity of carrier i on lane j was chosen from either $[0.3d_j, 0.5d_j]$ or $[0.8d_j, 1.0d_j]$. The smallest MQC that is implied by the first TQD interval, l_i^1, was randomly generated from $[0.1q_i, 0.5q_i]$, where $q_i = \sum_{j \in J} q_{ij}$ was the total capacity of the carrier. Set $u_i^k = l_i^1 + \frac{k}{3}(q_i - l_i^1)$ and $l_i^{k+1} = u_i^k + 1$ for $k \in K$. No discount was assigned to the first volume interval, i.e., $\alpha_i^1 = 0$. The discounts for the subsequent volume were increased with $\alpha_i^{k+1} = \alpha_i^k + \gamma$, where γ was the increment and randomly picked from {2%, 3%, 4%}.

All the generated data excluding discounts were rounded to integers. For each pair (m, n), a set of 10 instances was generated and tagged as "tptttqd-m-n". There are 90 instances in total.

3.4.2 Comparison results

In this section, we first test the importance of the initial upper bound (Section 3.4.2.1), which turns out to be critical to the performance of the BBB algorithm. Then, we investigate the efficiency improvement of the proposed bounding and branching strategies (Section 3.4.2.2). Finally, we evaluate the performance of our BBB algorithm by comparing with an improved i-constraint method [40] (Section 3.4.2.3).

3.4.2.1 Impact of initial upper bound set

We implemented the BBB methods with and without an initial upper bound set, and denoted the two implementations by BBB-UBS and BBB-no-UBS, respectively. The BBB-UBS adopts the BFS strategy and Fathoming Rules (1)—(3), while BBB-noUBS uses a depth-first search (DFS) strategy and only Fathoming Rules (1) and (2). Because the lexicographic minimal points of TPTT-TQD obtained by the initial upper bound set generating method are required by the Fathoming Rule (3) and BFS strategy. Eight small instances are randomly chosen as the test set.

We compared methods with the performance profile [29], which estimates the performance difference between methods. The performance profile $\rho(\tau)$ is the percentage of instances that are solved to the optimality by a method within a factor τ of the computational time required by the fastest one. A larger factor τ means a greater performance difference exists between the methods. In particular, $\rho(1)$ is the percentage of instances solved fastest by the method.

The performance profile of BBB-UBS and BBB-no-UBS is shown in Figure 3.5. The value $\rho(1)$ demonstrates that BBB-UBS is faster than BBB-no-UBS on all solved test instances. Moreover, BBB-UBS always outperforms BBB-no-UBS even with factor τ up to 10. It is obvious that more instances are solved to the optimality by BBB-UBS (87.5%). Overall, BBB-UBS is far more efficient than BBB-no-UBS. The initial upper bound set is used as default in the BBB algorithm for further tests.

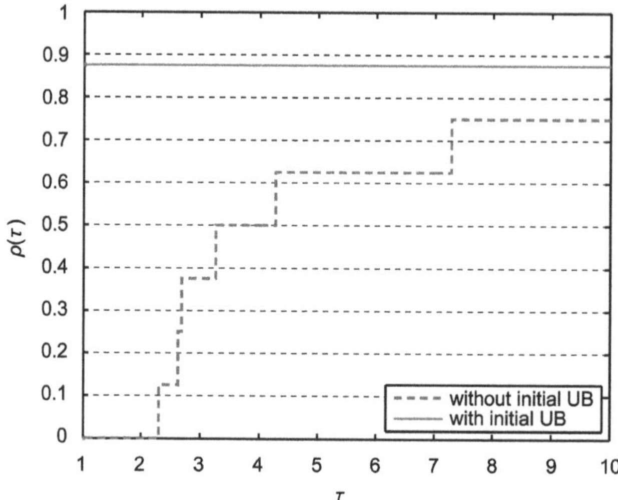

Figure 3.5 Performance profile of BBBs with and without initial upper bound set

3.4.2.2 Impacts of bounding and branching strategies

To evaluate the improvements of our two bounding strategies and the branching mechanism, the following four BBB variants were tested:

- BBB: the basic bi-objective branch-and-bound algorithm that implements the Fathoming Rules (1)—(3) and the initial upper bound set;
- BBB-SSR: the bi-objective branch-and-bound algorithm based on BBB that uses Bounding-SRR to implement the Fathoming Rule (4);
- BBB-SSR-FS: the bi-objective branch-and-bound algorithm based on BBB-SSR with the additional Bounding-FS to implement the Fathoming Rule (5);
- BBB-SSR-FS-HB: the bi-objective branch-and-bound algorithm based on BBBSSR-FS using the hybrid branching strategy.

Figure 3.6(a) shows the performance profile for BBB and BBB-SSR with τ taking values from [1, 10] on 8 randomly chosen small test instances. It shows that BBB-SSR was faster than BBB. BBB-SSR appears to perform far better than BBB because we see that the lines of BBB totally position underneath the lines of BBB-SRR. In particular, BBB-SRR solved all instances which are 12.5% more than BBB. The Bounding-SRR strategy yields a significant performance improvement.

Figure 3.6(b) depicts the effects of Bounding-FS strategy on the chosen test instances. From the performance profile, BBB-SSR-FS is faster to solve half of the instances while BBB-SSR performs better on the other half. However, before the factor τ extends to 1.06, BBB-SSR-FS stands out. Then, both BBB-SSR and BBB-SSR-FS can solve all test instances optimally if the factor τ is larger than 1.06. Overall, the performance profile shows that BBB-SSR-FS can slightly improve the performance.

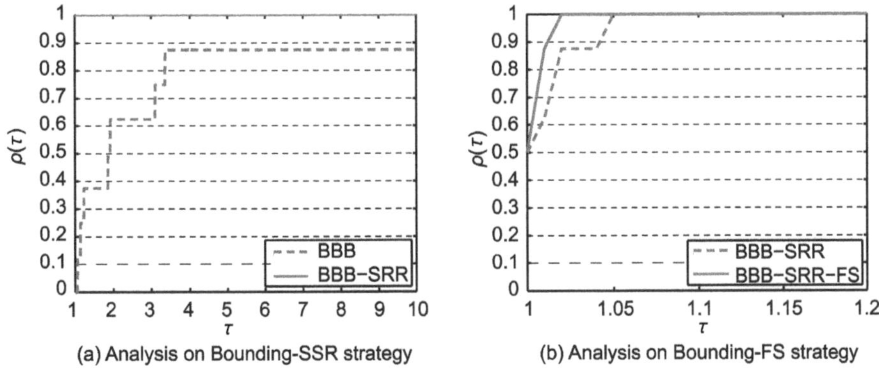

(a) Analysis on Bounding-SSR strategy

(b) Analysis on Bounding-FS strategy

Figure 3.6 Performance profiles on BBBs with two Bounding strategies

Figure 3.7 shows the performance profile for BBB-SSR-FS and BBB-SSR-FS-HB with τ in the interval [1, 2]. Clearly, BBB-SSR-FS-HB is faster than BBB-SSR-FS on all the test instances. This is because the Pareto branching decreases the average computational time spent on processing each node, though it might increase the number of nodes in the search tree. The proposed hybrid branching strategy has a positive effect on the overall performance.

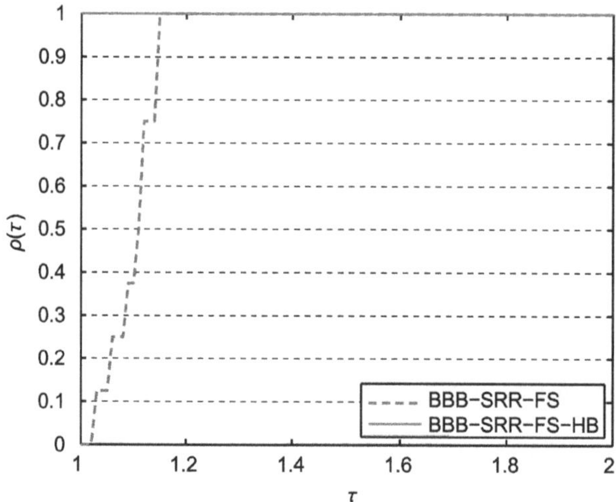

Figure 3.7 Performance profile of BBB with hybrid branching

Moreover, to compare the performance of the four variants on the instances that cannot be optimally solved, we constructed another test set consisting of 6 randomly chosen large size instances. We applied the four variants on this large test set and measured the obtained Pareto front using the hypervolume indicator. The solutions in the front were recorded every 1200 seconds for each BBB variant. As reported in Figure 3.8, it can be seen that BBB-SSR and BBB-SSR-FS-HB achieve a better approximation to the true Pareto front of TPTT-TQD within 3600 seconds. When the running time becomes more than 4800 seconds, BBB-SSR-FS and BBB-SSR-FS-HB outperform the other two variants.

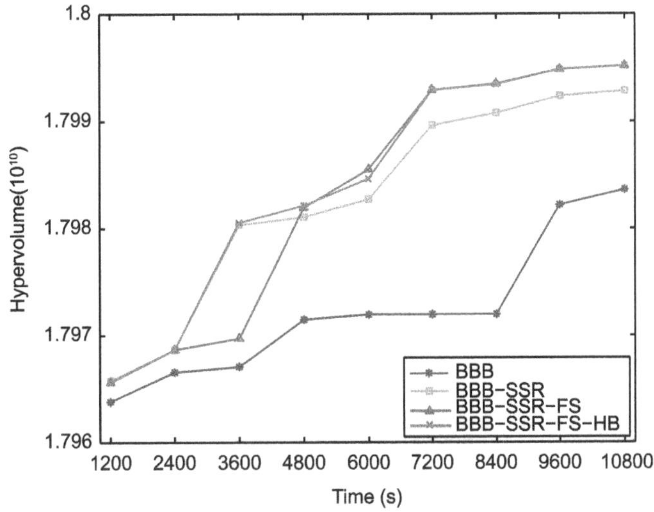

Figure 3.8 Hypervolume value on a large TPTT-TQD test set

In summary, BBB-SSR-FS-HB performs best because it solves more instances to optimality with less computational time. In addition, for those instances cannot be solved optimally, BBB-SSR-FS-HB provides a better approximation to the Pareto front. Consequently, BBB-SSR-FS-HB is our best choice among all implemented variants.

3.4.2.3 Comparison against an existing approach

In this section, we applied BBB-SSR-FS-HB to solve all the generated instances. The improved ϵ-constraint [40] which can compute evenly distributed nondominated points on the Pareto front was used to measure the performance of our BBB-SSRFS-HB method. The ϵ-constraint method produces a nondominated point using an IP solver, and iteratively compute more nondominated points by varying the upper bounds on the objectives. Since many IP programs are solved, the computational time is generally large. To conduct a fair comparison, for an instance that is optimally solved by BBB-SSR-FS-HB, its computational time was set as the time limit of ϵ-constraint method on this instance; otherwise, the time limit was set to three hours.

The detailed comparison results for all optimally solved instances are given in Table 3.1. The column *#Opt* presents the number of instances solved to the optimality. For each method, the columns *#ND* and *Time(s)* give the number of the nondominated points and the average computational time in seconds respectively, and the column *H* provides the average hypervolume value of the Pareto fronts.

Table 3.1 shows that our BBB-SSR-FS-HB method optimally solves 70 out of 90 TPTT-TQD instances, in which 30, 28, and 12 instances are solved with 10, 15, and 20 carriers, respectively. BBB-SSR-FS-HB performs significantly better than the ϵ-constraint method in terms of the number of the nondominated points and the hypervolume value for all instances.

Overall, our BBB-SSR-FS-HB is generally efficient to solve instances of practical size in the maritime transportation, where the number of prominent carriers in an auction is around 10. However, if the instance size reaches up to (20,400), BBB-SSR-FS-HB cannot solve them to optimality due to the dramatic increase of the number of nodes in the search tree, and then a multi-objective heuristic is more suitable since it is able to find several nondominated points.

Table 3.1 Comparison results on TPTT-TQD instances

Set	#inst	BBB-SSR-FS-HB				ϵ-constraint		
		#Opt	*#ND*	*H*(10^8)	*Time*(s)	*#ND*	*H*(10^8)	*Time*(s)
tptttqd-10-100	10	10	195	6.15	73.5	69	6.11	74.4
tptttqd-10-200	10	10	266	69.28	139.0	79	69.02	140.3
tptttqd-10-400	10	10	494	770.30	212.1	64	764.69	216.5
tptttqd-15-100	10	10	253	7.51	2115.6	183	7.51	2419.6
tptttqd-15-200	10	10	308	28.09	3853.5	219	27.77	3861.6
tptttqd-15-400	10	8	304	194.20	4884.1	144	193.79	4899.4
tptttqd-20-100	10	8	239	8.19	5134.1	211	8.14	5139.6
tptttqd-20-200	10	4	287	32.02	5912.1	198	31.91	5916.4
tptttqd-20-400	10	0	—	—	—	—	—	—

3.4.3 Insights

As introduced, carriers are requested to update MQCs and discounts in a new round of the auction and generally would also accept allocations where the old MQCs and discounts from previous rounds are satisfied. We construct two scenarios:

(A) solve the TPTT with only the MQCs and discounts received in the current round;

(B) solve the TPTT-TQD with total quantity structures that consist of TQDs from both the previous rounds and the current round.

By comparing the two scenarios, we answer whether introducing total quantity discounts and solving the TPTT-TQD can improve the decision of the shipper.

The two scenarios are simulated on a random instance "case-

10-100-2". In the instance, each carrier offers three pairs of MQC and discount. In other words, each carrier offers a TQD structure where three volume intervals exist. In scenario (A), three rounds of a procurement auction are executed and each carrier offers a larger discount after a round. The shipper solves the TPTT for each round and obtains three Pareto fronts, as shown in Figure 3.9(a). In scenario (B), the TPTT-TQD is solved once by the shipper and a Pareto front, as shown in Figure 3.9(b).

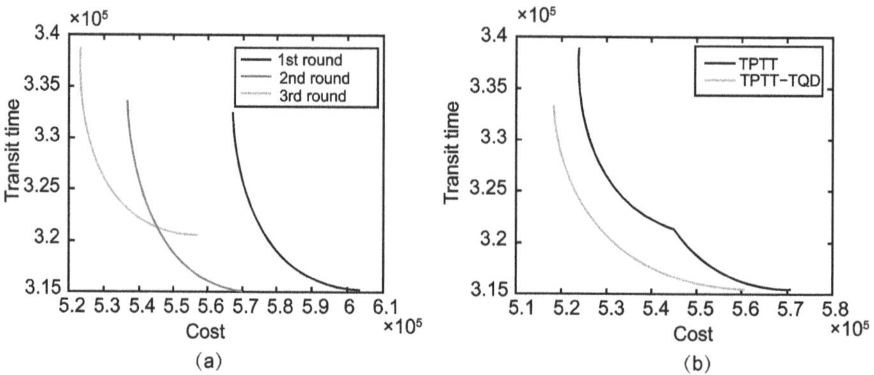

Figure 3.9 Simulation on instance "tptttqd-10-100-2"

In scenario (A), the shipper needs consider all the three fronts and finally decides the freight allocation plan based on a nondominated point. As shown in Figure 3.9(a), some points on the fronts of the second and third rounds are nondominated, which are also plotted in Figure 3.9(b) after discarding the dominated points. Figure 3.9(b) shows that the Pareto front obtained by solving the TPTT-TQD in scenario (B) dominates that of scenario (A). That is, by considering all the possible TQDs in a total quantity structure, the decision of the shipper can be improved by reducing both the total

cost and total transit time. It also reveals that the shipper might be motivated to request carriers to offer multiple TQDs to achieve more cost reduction and transit time reduction.

In summary, the total quantity discounts introduce more flexibility in the decision-making of the shipper. Though the complexity is larger due to more variables involved, the TPTT-TQD is worth to be solved so that the shipper's decision is improved. Moreover, the proposed BBB algorithm is applicable to compute a Pareto front consisting of ESN points for the TPTT-TQD especially for the applications where a small number of carriers participates in the auction, such as carriers in the sector of sea freight transportation.

3.5 Summary

In this chapter, we present a new model for the bi-objective transportation service procurement problem with both transit time and total quantity discounts. To obtain the Pareto front, a bi-objective branch-and-bound algorithm is designed, in which two new stronger fathoming rules, two novel bounding procedures with search region reduction, and a hybrid branching strategy are designed. By considering the total quantity discounts in the auction, we find some insights analytically that the new auction process has a strong advantage over the original one for the shipper to improve the decision. Computational experiments show that our method performs effectively, especially on small size instances. Future research can focus on designing more efficient strategies for the components of bi-objective branch-and-bound method to solve the large size instances. In addition, multi-objective heuristic algorithms can be designed and tested.

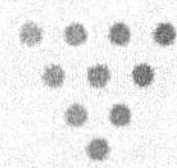

Chapter 4
Bi-objective Memetic Algorithm for the O2O Food Pickup and Delivery Problem with Time Windows

4.1 Introduction

With the rapid growth of internet technologies and logistics, online-to-offline (O2O) e-commerce becomes more and more popular. O2O platforms allow customers order goods (service) online, then the goods (service) will be delivered offline. In the takeout market, O2O takeout companies, such as Meituan and Ele.me, provide platforms to match food supplies and demands between restaurants and customers. Once an online order for some food of a restaurant is received from a customer, the O2O takeout company will assign the order to a rider to fulfill the offline pickup and delivery service; that is, the rider will pick up the food from the restaurant and deliver it to the customer. In China, the O2O takeout market continues to grow at a high rate, and is expected to exceed 3600 billion in 2018 [92].

In practice, an O2O takeout platform usually runs an optimization engine to seek solutions that allocate orders to riders and determine the delivery routing plans to serve all the orders so as to minimize the operating costs. Such an optimization problem computed by the O2O platform is an O2O food pickup and delivery problem, where a set of routes has to be constructed to fulfill the pickups and deliveries for the orders. In the problem, each order specifies the amount of food to be delivered, the location of the restaurant where it is to be picked up from and the location of the customer where it is to be delivered to. In addition, some practical constraints, e.g., time windows, the maximum working duration of a rider, and the capacities of vehicles, are needed to be considered. Moreover, a solution of the O2O food pickup and delivery problem consists of a set of "open" routes because such a route ends at the last serviced customer instead of the depot. The rider does not need to return to the depot after all the assigned orders are fulfilled.

Generally, the O2O takeout platform can assign an order to a rider with some classical criteria; for instance, the order can be first assigned to the rider that is nearest to the restaurant. Unfortunately, such a simple strategy may not be helpful in improving the pickup and delivery solution for the logistics system that deals with all the existing orders. Particularly, the strategy that always assigns orders to the nearest rider cannot guarantee to find a good solution that minimizes the total travel distance for the pickup and delivery problem. Since the classical pickup and delivery problem is NP-hard, the simple criteria might not construct good solutions all the time. Therefore, it is important to find optimized solutions to reduce the operating costs for the O2O takeout platform. We thus are motivated to study the possible concerns faced by the platform and attempt to develop optimization algorithms for their problem.

In the classical pickup and delivery problems, minimizing the total travel time or travel distance is quite common. However, such a single objective might not be enough for the O2O takeout platform. In the O2O takeout market, customer satisfaction is much more important for the success in the long-term intense competition with rivals. Many factors are related to customer satisfaction. Among them, prices, the quality of foods, and the quality of the delivery service are critical. Prices and the quality of foods depend on the restaurants registered on the platform. By including more restaurants into the platform, food diversity and price competition can be increased. But the delivery service is operated by the delivery teams of the platform and the service quality is determined by the routing solution. Once an order is placed, the customer who stays hungry always expects to receive his food as soon as possible. Thus, the delivery service should be responsive and rapid, and the delivery routing solution should reduce the waiting times of the customers.

Since food is perishable, the agile deliveries are also of great benefits to the freshness of the ordered food. Therefore, in addition to the operating costs that are most related to the total travel distance, the total waiting time of the customers is also most concerned by the O2O takeout platform.

We study a bi-objective O2O food pickup and delivery problem with time windows (O2O-PDPTW). It minimizes the total duration time of all riders and the total waiting time of all customers simultaneously, and decides the order allocations and delivery routing plans. If the goal of the O2O platform is only to save operating costs, it tries to fulfill the pickups and deliveries for the orders with a minimum number of riders. But if the goal is to make customers more satisfied, the platform needs to provide more riders to serve these orders. Therefore, operating costs and customer satisfaction are two conflicting objectives.

The contribution of this chapter is threefold. First, we introduce the bi-objective of O2O food pickup and delivery problem with time windows, and we provide a mathematical model for it. Second, we propose a bi-objective memetic algorithm in which a multi-objective evolutionary framework is combined with multi-directional local search (MDLS). Third, we evaluate the developed approach on a number of instances, and analyze the trade-offs in the decision-making procedure. Experiments show that our approach can approximate the Pareto front very well.

The remaining of the chapter is organized as follows. The formulation of the O2O-PDPTW is described in Section 4.2. The bi-objective memetic algorithm is presented in Section 4.3. Section 4.4 presents the computational experiments. Finally, in Section 4.5, we draw conclusions and suggest directions for future research.

4.2 Problem and definitions

We consider a pickup and delivery problem for the O2O delivery applications, i.e., the O2O food delivery case. In China, customers can order foods through mobile APPs operated by O2O companies. Such a company matches the food demands from customers and supplies from restaurants, and manages delivery teams to execute the food deliveries. Once a food order is received, the platform of the company will notify the restaurant to prepare food and assign the delivery request for the order to a rider. The rider then goes to pickup the food at the restaurant and deliver it to the customer.

Suppose there are n orders just received in the platform, and m riders are ready to leave the depot (operation center) to start deliveries. Let $R = \{1, \cdots, n\}$ include all the existing orders and $K = \{1, \cdots, m\}$ include all the riders. For order $i \in R$, let i and $n+i$ denote the location of the restaurant and the location of the customer, respectively. Let $N^+ = \{1, 2, \cdots, n\}$, $N^- = \{n+1, n+2, \cdots, 2n\}$, and $N = N^+ \cup N^-$. The graph $G = (V, A)$ consists of the nodes $V = N \cup \{0, 2n+1\}$ and the arcs $A = \{(0, i) : i \in N\} \cup \{(i, j) : i, j \in N\} \cup \{(i, 2n+1) : i \in N\}$, where 0 represents the depot and $2n+1$ is an artificial sink node. For each edge $(i, j) \in A$, we assign a travel time $t_{ij} \geqslant 0$ and the times satisfy the triangle inequality, i.e., $t_{il} + t_{lj} \geqslant t_{ij}$ for all $i, j, l \in N \cup \{0\}$. Particularly, we have $t_{i,2n+1} = 0$ for all $i \in N \cup \{0\}$.

Each order $i \in R$ has a time window $[a_i, b_i]$ and the delivery must be finished within the time window. That is, the food is ready for pickup after time a_i and must be delivered to the customer by time b_i. If the rider arrives early before the earliest pickup time a_i, he has to wait until the start of the time window. The service times at node $i \in N$ are s_i and there is $s_0 = 0$. Each rider can work within the time window $[0, T]$, where T represents the maximum working duration. Each rider leaves the depot at time 0. After the last customer is serviced, the delivery route ends and the rider returns home (node $2n + 1$) directly instead of to the depot. The duration time of the rider is thus equal to the time when he leaves the last customer.

The food quantity of order $i \in R$ is $p_i \geqslant 0$. At the nodes $i \in N^+$ and $n + i \in N^-$, the amounts of goods that must be loaded by a rider are $q_i = p_i$ and $q_{n+i} = -p_i$, respectively. The vehicles used for food delivery are electric bicycles. Each electric bicycle of a rider has the same capacity Q. When a rider leaves the depot, the electric bicycle is empty.

We introduce a mathematical model for the O2O-PDPTW, adapted from the model of the pickup and delivery problem proposed by Ropke and Pisinger [83]. For each arc $(i, j) \in A$ and each rider $k \in K$, let x_{ij}^k be a binary variable that equals 1 if and only if rider k travels directly from node i to node j. For each rider $k \in K$, let S_i^k be a nonnegative integer variable that indicates when rider k starts the service at node $i \in V$, and L_i^k be a nonnegative integer variable that indicates the amount of goods on the electric bicycle of rider k at node $i \in N$ before starting service. S_i^k and L_i^k are only defined when a rider k actually visits node i. For convenience, we summarize the notations used below:

Parameters

R the set of orders, $R = \{1, \cdots, n\}$;

K the set of riders, $K = \{1, \cdots, m\}$;

a_i the earliest pickup time of order i;

b_i the latest delivery time of order i;

q_i the food quantity at node i;

s_i the service time at node i;

t_{ij} the travel time between node i and node j;

T the maximum working duration of riders;

Q the capacity of a rider's electric bicycle;

Decision variables

x_{ij}^k a binary variable indicating whether rider $k \in K$ travels on arc $(i, j) \in A$;

S_i^k a nonnegative integer variable indicating when rider $k \in K$ starts the service at node $i \in V$;

Parameters

L_i^k a nonnegative variable indicating the amount of food on rider $k \in K$ before servicing node at $i \in N$.

The O2O-PDPTW is formulated in the following:

$$z_1 = \min \sum_{k \in K} S_{2n+1}^k \tag{4.1}$$

$$z_2 = \min \sum_{i \in N^-} \left(\sum_{k \in K} S_i^k - a_i\right) \tag{4.2}$$

$$\text{s.t.} \sum_{k \in K} \sum_{j \in N} x_{ij}^k = 1, \qquad \forall i \in N^+ \tag{4.3}$$

$$\sum_{j \in N} x_{ij}^k - \sum_{j \in N} x_{j,n+i}^k = 0, \qquad \forall i \in N^+, k \in K \tag{4.4}$$

$$\sum_{j \in N \cup \{2n+1\}} x_{ij}^k - \sum_{j \in N \cup \{0\}} x_{ji}^k = 0, \qquad \forall i \in N, k \in K \tag{4.5}$$

$$\sum_{i \in N \cup \{2n+1\}} x_{0i}^k = 1, \qquad \forall k \in K \qquad (4.6)$$

$$\sum_{i \in N \cup \{0\}} x_{i,2n+1}^k = 1, \qquad \forall k \in K \qquad (4.7)$$

$$S_i^k + s_i + t_{ij} \leqslant S_j^k + M(1 - x_{ij}^k), \qquad \forall (i, j) \in A, k \in K \qquad (4.8)$$

$$S_i^k \leqslant b_i, \qquad i \in N^-, k \in K \qquad (4.9)$$

$$S_i^k \geqslant a_i \sum_{j \in N} x_{ij}^k, \qquad i \in N^+, k \in K \qquad (4.10)$$

$$S_i^k \leqslant S_{n+i}^k, \qquad \forall i \in N^+, k \in K \qquad (4.11)$$

$$S_{2n+1}^k \leqslant T, \qquad \forall k \in K \qquad (4.12)$$

$$L_i^k + q_i \leqslant L_j^k + M(1 - x_{ij}^k), \qquad \forall (i, j) \in A, k \in K \qquad (4.13)$$

$$L_i^k \leqslant Q, \qquad \forall i \in N, k \in K \qquad (4.14)$$

$$L_0^k = L_{2n+1}^k = 0, \qquad \forall k \in K \qquad (4.15)$$

$$x_{ij}^k \in \{0, 1\}, \qquad \forall (i, j) \in A, k \in K \qquad (4.16)$$

$$S_i^k, L_i^k \in \mathbb{Z}^+, \qquad \forall i \in N, k \in K \qquad (4.17)$$

Objective (4.1) minimizes the total duration time of all the riders. Objective (4.2) minimizes the total waiting time of all the customers. The waiting time of a customer is equal to the time when the customer actually receives the food minus when the time the restaurant is ready for the food.

Constraints (4.3) state that each order must be served. Constraints (4.4) state that the food of each order is first picked up and then delivered by the same rider. Constraints (4.5) ensure flow balance on the network. Constraints (4.6) and constraints (4.7) ensure that each route starts at depot and ends at node $2n$+1.

The consistency of the time variables is ensured by constraints (4.8). The constraints also ensure that no subtours occur. The time window constraints are imposed by inequalities (4.9)—(4.10).

Constraints (4.11) ensure that each pickup occurs before the corresponding delivery. Constraints (4.12) state that the duration of any rider does not exceed the maximum working duration. The consistency of the load variables is ensured by constraints (4.13). Constraints (4.14)—(4.15) ensure that the load variables are set correctly along the routes. Finally, constraints (4.16)—(4.17) indicates the domain of the decision variables.

Due to the O2O-PDPTW being NP-hard, exact approaches can only solve relatively small instances. Therefore, we need to develop a multi-objective heuristic to solve the large instances of practical applications.

4.3 Solution method

We propose a bi-objective memetic algorithm (BMA) adapted from the memetic algorithm proposed by Moscato et al. [63] to solve the O2O-PDPTW. The memetic algorithm(MA) is a hybrid genetic algorithm based on the idea of using a population-based evolutionary algorithm to achieve diversity and local search methods to achiever intensity. The MA is now widely used to solve many optimization problems, such as scheduling problems [35, 50, 89], allocation problems [44, 98] and vehicle routing problems [19, 64, 67, 90, 105]. For a comprehensive review on memetic algorithms, the readers can refer to the survey paper by Neri et al.[66].

Inspired by the idea of MA, we design a BMA that uses the multi-directional local search (MDLS) strategy embedded in a multi-objective evolutionary framework. MDLS [91] is an efficient heuristic for the multi-objective optimization problems. The main idea of MDLS is that it searches for the neighbors of the solution in the direction of only one objective at a time.

A solution of the O2O-PDPTW can be represented as a set of feasible routes (see Section 4.3.1 and Section 4.3.2). The BMA proceeds as follows. At the beginning, it starts to construct an evenly distributed initial population using a two-phase constructive algorithm (see Section 4.3.3), and then initialize the feasible route pool Ω. In each iteration, the BMA iteratively performs four-phases: crossover, selection, MDLS and a route combination. First, the crossover operator (see Section 4.3.4) is invoked, and all the newly generated solutions are added to the population. Among all the solutions in the population, only *N* solutions can be survived in the selection procedure of NSGA-II algorithm [27]. Subsequently, we try to search the neighbors of the solutions using the adapted MDLS strategy (see Section 4.3.5). To save the computation time, MDLS is only performed on the nondominated solutions instead of the whole population. Finally, a route combination method (see Section 4.3.6) is applied on the pool Ω to obtain a high-quality solution. Note that the pool Ω is maintained and updated when new solutions are discovered.

The BMA algorithm stops when the maximum iteration number or the time limit is reached. Then the nondominated solution set forms an approximation of the Pareto front. The pseudocode for BMA is shown in Algorithm 4.1.

Algorithm 4.1 Bi-objective memetic algorithm(BMA)

```
BMA(N, MaxIter, TimeLimit)
Ω = Ø, population = Ø, iter = 0;
Generate N initial solutions and add to population;
Add feasible routes to Ω;
while iter ⩽ MaxIter and TimeLimit is not reached
    Generate N solutions using crossover operators;
    Add new solutions to population;
    Select N solutions among population;
    Generate 2N solutions using the MDLS and add to population;
```

Continued

Update Ω;

Generate a solution using route combination and add to *population*;

iter = *iter* + 1;

return the nondominated solutions in *population*.

4.3.1 Solution representation

A solution of the O2O-PDPTW is a set of "open" routes. Each route can be represented as a sequence of pickup and delivery nodes that starts at the depot and terminates at an artificial sink node. The route is feasible only if it satisfies the time windows, the capacity, the maximal duration limit, the pairing and the precedence constraints. A solution is feasible if every route is feasible and all orders are served exactly once.

4.3.2 Route feasibility

Since the feasibility check of the routes will be frequently used in the main components of the BMA, we consider to identify a set of unreachable arcs in the preprocessing procedure and store them in a matrix. Then, before inserting a node into a route or removing a node from a route, we first check whether the new arcs are reachable. If one of them is unreachable, the insertion or removal attempt is not allowed. The preprocessing procedures are helpful in rejecting infeasible routes.

Because of the time windows, the capacity and the paring constraints, several arcs in fact can be eliminated from the reachable arc matrix. The preprocessing steps can be defined as follows:

- arcs $(n + i, i)$, $(0, n + i)$, $(i, 2n + 1)$ are infeasible for $i \in N^+$;
- arcs $(i, 0)$ and $(2n + 1, i)$ are infeasible for $i \in N$;
- arc (i, j) is infeasible if $a_i + s_i + t_{ij} > b_j$ for $i, j \in N$;
- arc (i, j) is infeasible if $q_i + q_j > Q$ for $i, j \in N^+$.

If a route consists of all reachable arcs, we adapt an eight step evaluation scheme [23] to examine the feasibility. It is based on the forward time slack introduced by savelsbergh [84]. The forward time slack defines the maximum amount of time by which the start time can be delayed at one node without causing the other nodes to violate the time windows and the maximum working duration constraints [42, 72, 105]. In addition, the route is further checked whether it violates the capacity constraints. That is, the amount of food at each node should be calculated.

4.3.3 Initial population

We develop a two-phase constructive algorithm to produce a set of evenly distributed solutions as the initial population for BMA, which helps to enhance the diversity of the population.

Given a weight vector (λ_1, λ_2) where $\lambda_1 + \lambda_2 = 1$ and $\lambda_1, \lambda_2 \geqslant 0$, let z_1 and z_2 be the total duration time and the total waiting time of all the used routes, the weighted-sum objective function can be defined as follows:

$$z_\lambda = \lambda_1 \cdot z_1 + \lambda_2 \cdot z_2$$

For a given weight vector $\lambda = (\lambda_1, \lambda_2)$, a feasible solution can be computed by the classic greedy insertion heuristic [83]. First, all orders, also called pickup-delivery (PD) pairs, are put into a list in a random order. Across all the routes, a PD pair insertion is tried at every position of each route. Whenever an unserved PD pair can be inserted in a route, we record its increment in the weighted-sum objective function and two insert positions. After attempting to insert the PD pair in all the used routes, the greedy insertion heuristic chooses the best route with the smallest increment in the weighted-sum objective function, and inserts the PD pair at the best positions of the route. If the PD pair cannot be inserted into any route, a new route is used. This process continues until all the PD pairs are inserted. Finally, a new feasible solution is obtained.

In the two-phase constructive algorithm, the first phase is to find two feasible solutions close to the corner points on the true Pareto front of O2O-PDPTW. Therefore, the greedy insert heuristic first performs on weight vector (1, 0) and (0, 1). In the second phase, we obtain feasible solutions by setting different weight vectors, where λ_1 is randomly generated in [0, 1] and $\lambda_2 = 1 - \lambda_1$. The process continues until the size of the population reaches N.

4.3.4 Crossover operator

To enhance the diversity of the population, we employ a customized crossover operator [105]. The crossover operator selects two parent solutions and yields a child solution. Parents selection is performed through a binary tournament, which randomly selects two solutions from the population and keeps the one with a smaller sum of the objective (4.1) and objective (4.2).

Then, each route in parent P_1 is randomly chosen with a probability of 50% and inherited by its child solution. For the routes of parent P_2, the orders already in the child are removed, which ensures that each order is served at most once. The child solution also inherit the routes of parent P_2 with a probability of 50%. Finally, all the unserved orders are reinserted into the child solution by the greedy insertion heuristic, where the weight vector λ is generated randomly.

4.3.5 Multi-direction local search

The MDLS searches for the neighbors of the current solution in the direction of only one objective at a time. Tricoire [91] has been successful applied it to three different multi-objective optimization problems: the multi-objective multi-dimensional knapsack problem, the bi-objective set packing problem and the bi-objective orienteering problem. It is also shown that the MDLS is able to provide competitive results for the bi-objective home care scheduling problem [14], the bi-objective team orienteering problem with time windows [74], and the bi-objective dial-a-ride problem in patient transportation [62].

To drive the search towards intensification and diversification simultaneously, we adapt the MDLS with several acceptance criteria (Section 4.3.5.1) and several local search procedures (Section 4.3.5.2).

4.3.5.1 Acceptance criteria

The acceptance criterion allows the search to move from the current solution to a neighbor solution. In this section, we introduce three acceptance criteria from strong to weak. The Acceptance Criterion (1), extended from the acceptance criterion in single objective heuristics, can drive the search more intensification. However, some promising search regions are unable to be discovered. The Acceptance Criterion (2), which is developed by Tricoire [91], can enlarge the search region. However, the new neighbor solutricoiretions may improve one objective a little but deteriorate the other objective very much. Therefore, we design a new Acceptance Criterion (3) that combines the advantages of broadening the search region and improving the population.

Acceptance Criterion (1) For the current solution s, a new neighbor solution s' is accepted if s' dominates s.

Acceptance Criterion (2) For the current solution s, a new neighbor solution s' is accepted if s' is a comparable solution which improves one of the objectives.

Acceptance Criterion (3) For the current solution s, a new neighbor solution s' is accepted if $s'_1 = s_1 \times (1 - r\%)$ and $s'_2 \leqslant s_2 \times$

$(1 + r\%)$, or $s_2' = s_2 \times (1 - r\%)$ and $s_1' \leqslant s_1 \times (1 + r\%)$.

An illustration of the Acceptance Criterion (3) is shown in figure 4.1. The search region of a neighborhood solution is smaller than that of the Acceptance Criterion (2), but all the promising regions are retained. In addition, the parameter r can be calculated and updated dynamically throughout the search, which relieves us from the burden of the parameter tuning.

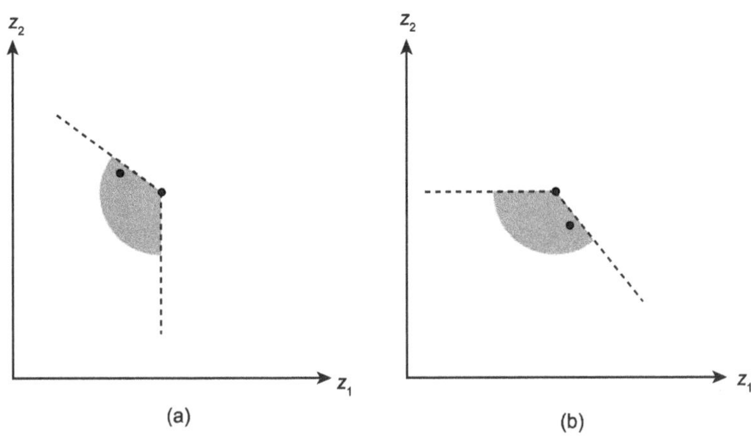

Figure 4.1 An illustration of Acceptance Criterion (3) in MDLS

The acceptance criterion is adjusted dynamically according to the Pareto front during the iteration. Due to the lack of diversity in the Acceptance Criterion (1), the MDLS is unlikely to find a good approximation by adopting the Acceptance Criteria (1), as observed in the preliminary experiments. Therefore, the Acceptance Criterion (3) is applied first. If the current Pareto front cannot be improved N_{no_imp} times, the Acceptance Criterion (2) is invoked to further enlarge the search region. We set $N_{no\_imp} = 20$ in the BMA implementations.

4.3.5.2 Local search procedures

Local search procedures are sequentially performed to explore the neighbor solutions.

- *inter-route relocate*: it iteratively removes a PD pair $[x^+, x^-]$ from one route and inserts it into another route.
- *exchange*: it iteratively selects two PD pairs $[x^+, x^-]$ and $[y^+, y^-]$ from routes A and B, and swaps them by inserting $[x^+, x^-]$ into route B and $[y^+, y^-]$ into route A. The insertion positions of the origin and the destination are unchanged.
- *2-opt*: it first removes two edges from two different routes. These two edges separate the two routes into four parts. If no order is split in two parts, the 2-opt operator is invoked to recombine them into two new feasible routes.

4.3.6 Route combination method

Route combination method uses the feasible routes in the pool Ω and recombines them to obtain solutions with good quality. First we formulate the O2O-PDPTW into a set partitioning model.

Let x_p be a binary variable that is equal to 1 if and only if the route p is selected in the solution. For each route $p \in \Omega$, let t_p be the duration time of the route and w_p be the waiting time of the route. Let b_{ip} be a binary variable that is equal to 1 if order i is served by p and 0 otherwise.

The route combination problem (RCP) can be formulated as

follows:

$$(\text{RCP}) \quad \min \sum_{p \in \Omega} (\lambda_1 t_p + \lambda_2 w_p) x_p \tag{4.18}$$

$$\sum_{p \in \Omega} b_{ip} x_p = 1, \quad \forall i \in N^+ \tag{4.19}$$

$$x_p \in \{0, 1\}, \quad \forall p \in \Omega^+ \tag{4.20}$$

The objective function (4.18) minimizes the weighted-sum of the total duration time and the total waiting time. Constraints (4.19) ensure that each order is served exactly once by one route. Constraints (4.20) impose binary integer constraints on x_p.

If we replace (4.20) by:

$$x_p \geqslant 0, \forall p \in \Omega \tag{4.21}$$

We obtain the linear relaxation of RCP. Given a weight vector λ, the nondominated solution can be found if the pool Ω includes all the feasible routes. Since the number of all possible feasible routes could be exponential, the RCP cannot be solved efficiently. In this work, the pool Ω is limited and the pool size N_{pool} is set too 500. With the pool Ω, an integer programming solver, such as ILOG CPLEX, is applied to solve RCP. The obtained solution is a high-quality feasible solution of O2O-PDPTW.

To accelerate the route combination procedure, we first obtain the lower bound by solving the linear relaxation of RCP on the pool Ω. If the lower bound is better than the weighted-sum objective z_λ of any nondominated solution in the population, then an integer programming solver is further applied.

4.4 Computational results

All algorithms were implemented as a sequential program in Java. The reported computational results were obtained by a PC equipped with an Intel(R) Core(TM) i74790 CPU clocked at 3.60 gigahertz, with 16 gigabytes RAM, and running Windows 8.1 Enterprise. The integer programming solver used was ILOG CPLEX 12.51.

After parameter tuning in preliminary tests, we decide the following parameter settings:

$N = 100$, $MaxIter = 100$

4.4.1 Test instances

The test instances are designed according to the characteristics of real-world O2O food takeout application. The generation procedure is similar to the one described by Lu et al. [52].

In the practical application, restaurants have more orders to fulfill in certain periods, such as meal time, and fewer orders to fulfill in other periods. For each test instance, we consider that the number of orders is multiple of the number of restaurants. The number of restaurants q was chosen from {6, 8, 12, 16, 20, 24}, and the number of orders n was set to δq, where δ was chosen from {1, 2, 4, 8}. Based on different δ, we divided the test instances into four categories, namely "A", "B", "C" and "D".

An instance set is named "O2O-PDPTW-g-q-n-T_{tw}", where g is the instance category and T_{tw} is the time window type from {1, 2, 3}.

The riders are assumed to be able to serve all orders, so the delivery team will have n available riders. We had a scheduling T = 4 (hours) horizon for all instances. Thus the time window for each rider was set to [0, 240] and the capacity Q was set to 20. The load q_i at restaurant i was chosen from the intervals [1, 2], [3, 6] or [7, 10] with a probability 0.7, 0.25 and 0.05, respectively, and the corresponding customer load $q_{i+n} = -q_i$. The service time s_i at each node i was chosen from the interval [1, 5]. Depending on different time window types T_{tw}, the time window width w_i for each order i was randomly selected from intervals [30, 60], [60, 120] or [90, 180]. The time window for order i was set to $[a_i, b_i]$, where a_i was randomly selected from the interval $[0, T - s_i - t_{i,n+i}]$ and $b_i = a_i + w_i$.

For all instances, the coordinates of each restaurant and customer location were chosen randomly, according to a uniform distribution over the [0,20] × [0,20] square. The depot was located at the middle of a square, namely [10,10].

For each set "O2O-PDPTW-g-q-n-T_{tw}", we randomly generated 10 instances. Therefore, a total of 720 instances are generated.

4.4.2 Performance measure

Recall that the well-known hypervolume indicator H [107] can be used to evaluate the performance of the bi-objective optimization methods.

Given an approximation set of points *A* and a reference set *B* in the objective space, the hypervolume *H*(*A*) represents the volume of the region that is bounded by a reference point, which is the nadir point of the union of set *A* and set *B*. With the hypervolume indicator *H*, larger value indicates better.

We tested two BMA implementations evaluate the route combination method. The basic implementation (BMA1) was the bi-objective memetic algorithm using the Acceptance Criterias (2) and (3). Based on BMA1, BMA2 was extended to use the route combination method.

To evaluate the performance of the bi-objective memetic algorithm, we adapted the NSGA-II [27] to solve all the test instances of O2O-PDPTW. In the implementation of NSGA-II, it uses the same initial population constructive algorithm and the customized crossover operator. However, it differs from BMA implementations in that it does not use the MDLS strategy in the multi-objective evolutionary framework.

The time limit is set to 300 seconds for each run of each BMA implementation. In order to ensure a fair comparison, the CPU time limit of NSGA-II is set to the actual computational time of BMA1 for each test instance.

4.4.3 Understanding the trade-off

In order to provide insights on the trade-off between the total duration time and the total waiting time, BMA1 was tested on a small instance, “O2O-PDP-A-6-6-3”. Figure 4.2 shows the Pareto front of instance "O2O-PDP-A-6-6-3" and the number of used riders for each solution on the Pareto front. As expected, solutions with shorter duration times require fewer riders to serve all orders, but this will result in longer waiting times. On the contrary, solutions with shorter waiting times require more riders to serve all orders, but duration times will increase.

Six nondominated solutions on the Pareto front of the instance is depicted in Figure 4.3. Each sub-figure represents the routing plan of a solution. Each route starts at the depot 0. The food of each order $i \in R$ is first picked up from node i^+ and then delivered to node i^{-1}.

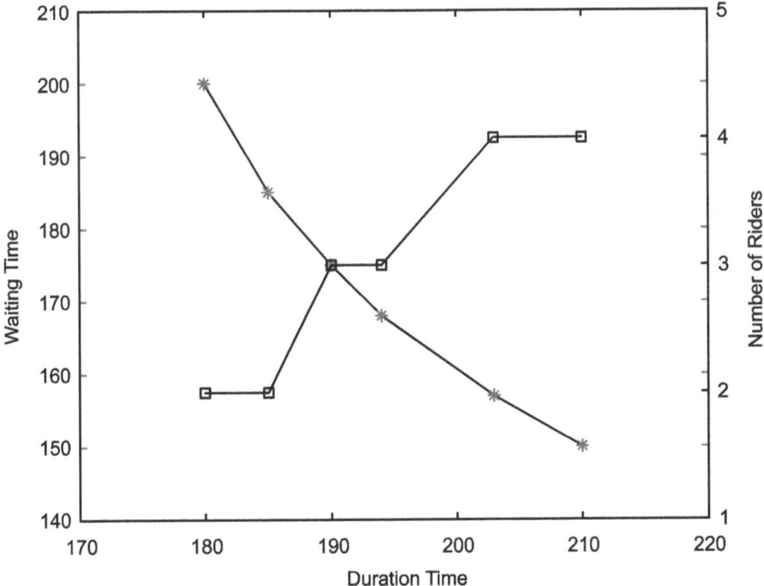

Figure 4.2 Trade-off of instance "O2O-PDP-A-6-6-3"

Solution 4.3(a) can achieve the minimum total duration time of 180 minutes by using two riders. Solution 4.3(b) uses the same number of riders but the routes are different, resulting in a longer total duration time and a shorter total waiting time.

The route in solution 4.3(b) is decomposed into two routes {0, 2^+, 4^+, 2^-, 4^-, 7} and {0, 6^+, 6^-, 5^+, 5^-, 7} in solution 4.3(c). The total duration time increases from 185 minutes to 190 minutes and the total waiting time decreases from 185 minutes to 175 minutes. Solution 4.3(d) differs from 4.3(c) in relocating order [3^+, 3^-] from one route to another, but the number of riders unchanged. We can also see that the last two solution 4.3(e) and 4.3(f) using four riders are in fact similar. In solution 4.3(f), customer node 2^- is relocated before the restaurant node 4^+, which can achieve the minimal total waiting time of 150 minutes.

Depending on the preference of the O2O takeout company, different solutions would be chosen. For example, if the decision maker prefers a minimal operating cost, he would choose the solution with total 190 minutes' duration time by using two riders, but the customers will wait about 30 minutes. If the decision maker aims at improving customer satisfaction, he would choose a solution with an average waiting time of about 25 minutes, but with 210 minutes duration time by using three riders. If the decision maker's preference is more complex, he still has other solutions on the Pareto front to choose from.

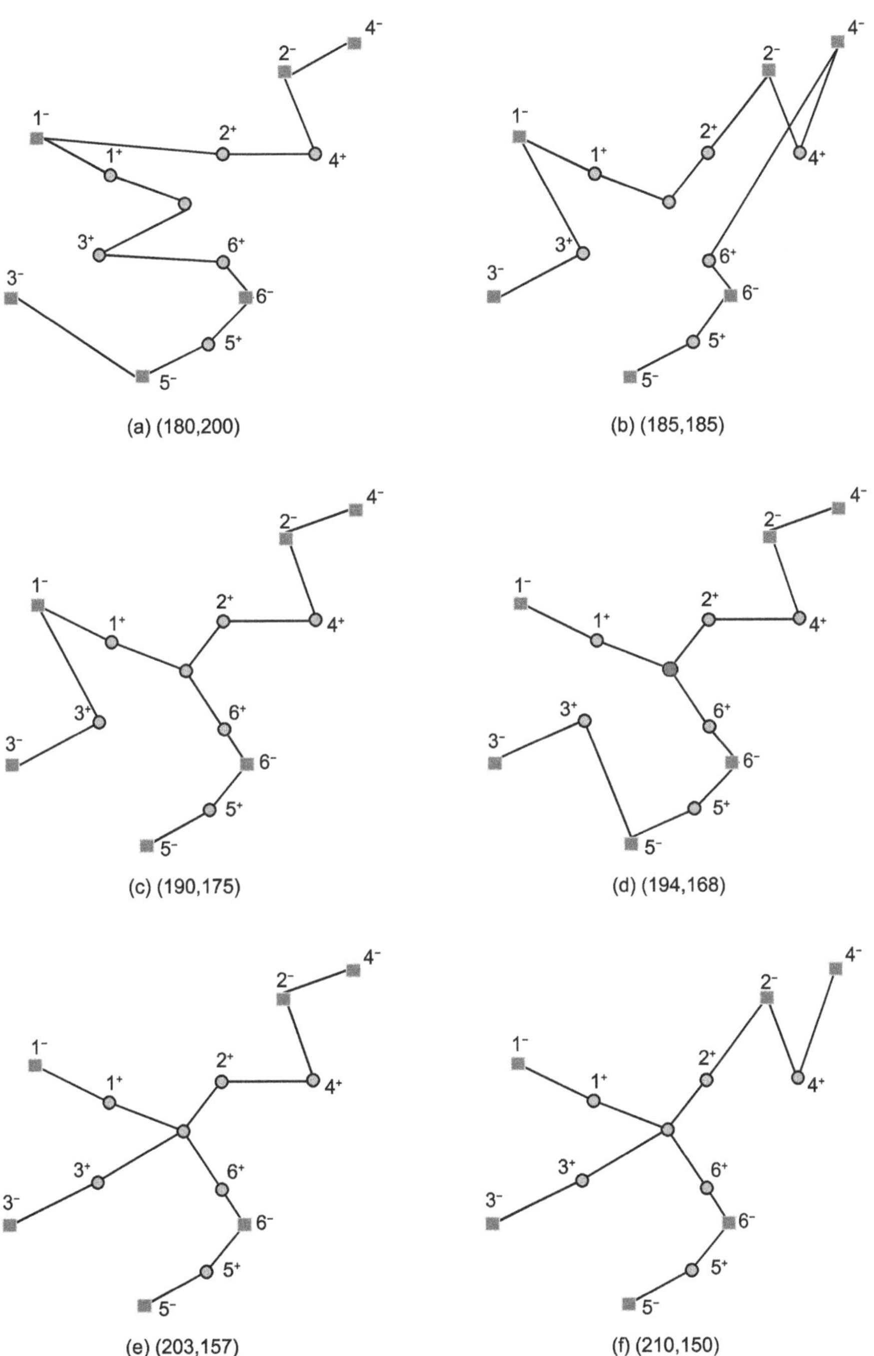

(a) (180,200)

(b) (185,185)

(c) (190,175)

(d) (194,168)

(e) (203,157)

(f) (210,150)

Figure 4.3 Routing plans of instance “O2O-PDP-A-6-6-3”

4.4.4 Effect of different time window widths

In this section, the effect of different time window widths on the Pareto front of O2O-PDPTW is investigated.

Figure 4.4 depicts the Pareto fronts generated by BMA1 for three instances: "O2O-PDP-A-12-12-1-8", "O2O-PDP-A-12-12-2-8" and "O2O-PDP-A-12-12-3-8". The instances are all the same except T_{tw} is different. We observed that the Pareto front of "O2O-PDP-A-12-12-3-8" was superior to the other two Pareto fronts, and some nondominated solutions were discovered in new regions of the objective space. It reveals that the O2O takeout platform might be motivated to choose the broader time windows to achieve more duration time reduction. However, if decision makers tend to make customers more satisfied, using the broad time windows will not improve decisionmaking.

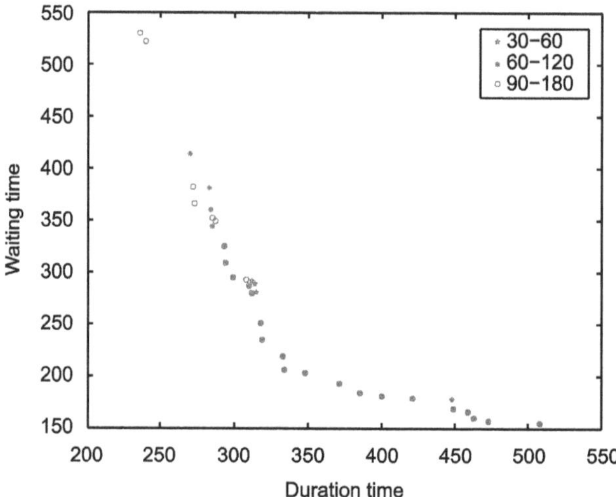

Figure 4.4 Effect of different time window widths on category A instances

Figure 4.5 depicts the Pareto fronts of instances "O2O-PDP-B-6-24-1-4", "O2OPDP-B-6-24-2-4" and "O2O-PDP-B-6-24-3-4". Although the nondominated solutions are distributed in a broader region in the objective space when T_{tw} = 3, it seems that some of them are dominated by other solutions on the other two fronts.

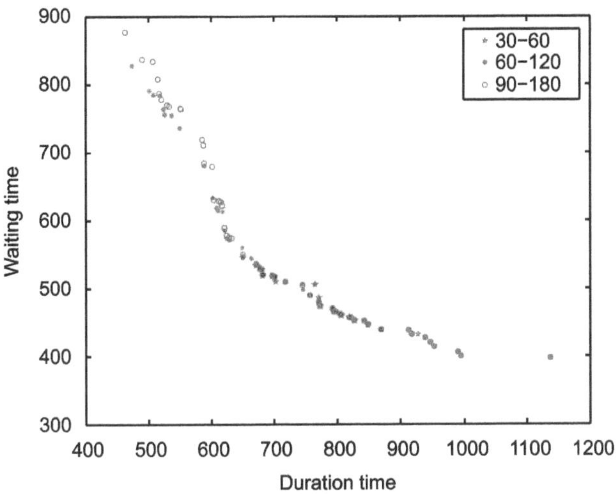

Figure 4.5 Effect of different time window widths on category B instances

In summary, there is a trade-off between the quality of the Pareto front and different time window widths. Generally, broader time windows introduce more flexibility in the decision-making of the O2O takeout platform. However, some solutions on the Pareto front may be worse.

4.4.5 Comparison against an existing approach

The detailed comparison results for 720 instances are given in

Table 4.1—Table 4.4. The column *#Solved* presents the number of instances solved. For each method, the columns *#ND* and *Time(s)* give the number of the nondominated points and the average computational time in seconds respectively, and the column *H* provides the average hypervolume value of the Pareto front.

Table 4.1—Table 4.4 shows that both BMA1 and BMA2 perform significantly better than the NSGA-II method in terms of the number of the nondominated points and the hypervolume value for all instances. To analyze the effect of the route combination method, we compare the performance of two BMA implementations. It shows that BMA1 has a relatively better performance in average than BMA2 in terms of indicator *H*. Due to the complexity of the route combination problem, it is very time-consuming to solve it. As a result, BMA2 in unable to improve the performance of BMA1.

BMA1 was applied on all instances. The detailed comparison results for instances in four categories are given in Table 4.5. The column *#restaurant* and column T_{tw} present the number of restaurants and the window time type. For each combination of #restaurant, T_{tw} and the instance category, the columns *#rider* give the average number of the riders of all the nondominated solutions. The columns *DT* and *WT* provide the average duration time and the average waiting time in minutes of all the nondominated solutions, respectively.

As revealed in Table 4.5, the order quantity has a large impact on the number of riders, the total duration time and the total waiting time. When the number of orders increases, they will all increase dramatically. On the contrary, the number of restaurants is not a critical

Table 4.1 Performance comparison of NSGA-II and BMA on O2O-PDPTW-A instances

Set	#Solved	BMA1			BMA2			NSGA-II		
		Time(s)	#ND	*H*	*Time*(s)	#ND	*H*	*Time*(s)	#ND	*H*
O2O-PDP-A-6-6-1	10	0.02	2.8	0.00	0.02	2.8	0.00	0.02	2.8	0.00
O2O-PDP-A-6-6-2	10	0.01	3.1	0.00	0.02	3.1	0.00	0.01	3.1	0.00
O2O-PDP-A-6-6-3	10	0.01	3.3	0.00	0.02	3.3	0.00	0.01	3.3	0.00
O2O-PDP-A-8-8-1	10	0.02	5.9	0.00	0.03	5.9	0.00	0.02	5.9	0.00
O2O-PDP-A-8-8-2	10	0.03	7.9	0.00	0.03	7.9	0.00	0.02	7.9	0.00
O2O-PDP-A-8-8-3	10	0.04	7.9	0.00	0.05	7.9	0.00	0.04	7.9	0.00
O2O-PDP-A-12-12-1	10	1.34	11.2	**18997.30**	1.59	11.2	18917.90	1.35	11.2	18280.80
O2O-PDP-A-12-12-2	10	1.66	16.5	65499.00	1.78	17.0	**65577.90**	1.67	15.1	64853.00
O2O-PDP-A-12-12-3	10	1.81	16.8	**73220.80**	1.93	16.4	73005.90	1.81	14.6	67814.10
O2O-PDP-A-16-16-1	10	1.79	23.7	**50264**	2.24	23.1	47429.3	1.80	21.4	49787.6
O2O-PDP-A-16-16-2	10	3.16	27.2	**79161.7**	3.45	27.6	77523.9	3.17	23.4	76827
O2O-PDP-A-16-16-3	10	4.15	26.2	**110534.1**	4.37	27.0	110428.1	4.16	22.8	103184.1
O2O-PDP-A-20-20-1	10	3.07	30.8	**79208.5**	3.72	29.9	78486.7	3.09	25.0	78797.1
O2O-PDP-A-20-20-2	10	5.84	34.9	158765.7	6.27	34.5	**159906.6**	5.87	27.4	153635.9
O2O-PDP-A-20-20-3	10	7.46	30.2	164835.4	7.96	32.5	**171707.2**	7.48	25.9	153380.4
O2O-PDP-A-24-24-1	10	4.39	30.8	108843.6	5.45	33.2	**119841.8**	4.41	26.1	107523.1
O2O-PDP-A-24-24-2	10	8.03	34.2	211606.2	8.76	31.9	**221643.1**	8.04	27.1	206989.3
O2O-PDP-A-24-24-3	10	10.82	31.8	**222540.7**	11.40	33.4	214405.4	10.85	27.2	208891.4

factor to bring in complexity. Although some orders are taken from the same restaurant, it is difficult to merge them because each order has a different time window constraint.

It is also shown in Table 4.5 that the time window widths have no impact on the number of riders. However, with the wide time windows, the average duration of the nondominated solutions becomes shorter and the average waiting time becomes longer.

Table 4.2 Performance comparison of NSGA-II and BMA on O2O-PDPTW-B instances

Set	#Solved	BMA1			BMA2			NSGA-II		
		Time(*s*)	#ND	*H*	*Time*(*s*)	#ND	*H*	*Time*(*s*)	#ND	*H*
O2O-PDP-B-6-12-1	10	1.26	11.9	**15745.60**	1.18	11.6	15738.60	1.27	11.3	15661.60
O2O-PDP-B-6-12-2	10	1.32	13.5	**40964.90**	1.56	13.8	40943.50	1.33	12.2	40477.00
O2O-PDP-B-6-12-3	10	1.75	13.5	42078.50	1.73	13.7	**42103.30**	1.75	11.9	41509.90
O2O-PDP-B-8-16-1	10	1.59	16.4	33762.30	1.96	16.7	**34401.90**	1.61	14.7	34799.10
O2O-PDP-B-8-16-2	10	2.53	17.7	56335.90	2.84	17.6	**56403.30**	2.54	16.3	55878.40
O2O-PDP-B-8-16-3	10	3.13	16.9	**52319.10**	3.48	16.5	52285.60	3.14	13.6	51343.00
O2O-PDP-B-12-24-1	10	5.12	40.0	120410.1	5.92	39.5	**126403**	5.14	33.0	119781.8
O2O-PDP-B-12-24-2	10	9.01	42.0	**265466.3**	9.62	43.1	248513.7	9.04	33.4	242095.6
O2O-PDP-B-12-24-3	10	12.46	39.2	**266431.6**	13.35	37.7	266251.7	12.49	33.3	251201.9
O2O-PDP-B-16-32-1	10	10.55	44.5	**227854.9**	12.09	46.6	226187.7	10.58	36.4	226416
O2O-PDP-B-16-32-2	10	19.68	46.1	**488470.7**	20.73	48.2	479625.2	19.72	34.5	457479.2
O2O-PDP-B-16-32-3	10	28.33	42.3	**535070**	29.41	43.9	534304.5	28.36	35.4	468998.4
O2O-PDP-B-20-40-1	10	18.77	57.7	**334845.1**	21.09	59.3	333611.6	18.81	46.5	318562.6
O2O-PDP-B-20-40-2	10	37.92	51.4	613414.4	40.15	49.2	**619437.2**	37.98	40.2	592472.9
O2O-PDP-B-20-40-3	10	51.05	47.7	**717844.9**	50.83	48.3	714244.8	51.08	38.7	612209.8
O2O-PDP-B-24-48-1	10	33.55	47.6	408824.2	36.85	48.8	**416622.2**	33.60	43.0	390251.1
O2O-PDP-B-24-48-2	10	60.74	44.2	**724895.4**	63.70	45.1	717021	60.79	39.0	629533.1
O2O-PDP-B-24-48-3	10	85.32	43.5	**707963.8**	87.67	41.1	681105.3	85.41	28.8	633438.1

Table 4.3 Performance comparison of NSGA-II and BMA on O2O-PDPTW-C instances

Set	#Solved	BMA1			BMA2			NSGA-II		
		Time(s)	#ND	*H*	*Time(s)*	#ND	*H*	*Time(s)*	#ND	*H*
O2O-PDP-C-6-24-1	10	4.42	32.6	129275.9	5.35	32.8	**130078.9**	4.43	29.8	126483.8
O2O-PDP-C-6-24-2	10	8.17	36.3	203920.8	8.95	36.7	**206205.8**	8.20	30.1	203997.1
O2O-PDP-C-6-24-3	10	11.22	34.0	217072.1	12.10	33.9	**223158.2**	11.24	24.6	197855.5
O2O-PDP-C-8-32-1	10	10.16	49.9	236097.9	11.33	48.6	**240730.9**	10.19	42.7	236492
O2O-PDP-C-8-32-2	10	19.04	50.7	476352.7	20.86	46.8	478551.6	19.08	40.5	458841.7
O2O-PDP-C-8-32-3	10	26.53	47.5	513749.9	27.94	45.2	**511286.1**	26.57	38.4	470634.1
O2O-PDP-C-12-48-1	10	33.69	50.7	547575.8	37.06	53.1	**550536.9**	33.74	50.0	516981.3
O2O-PDP-C-12-48-2	10	63.54	49.2	**903010.9**	65.66	51.9	898197.6	63.62	46.1	854017.1
O2O-PDP-C-12-48-3	10	88.16	45.3	879503.1	93.29	49.5	**884079**	88.25	37.7	875193.8
O2O-PDP-C-16-64-1	10	83.48	60.0	907975.8	91.40	59.2	**941989.9**	83.56	56.5	846402.1
O2O-PDP-C-16-64-2	10	161.31	56.5	**1365602**	162.34	53.8	1359701	161.39	63.7	1198368
O2O-PDP-C-16-64-3	10	225.86	55.8	**1543120**	229.54	55.2	1543118	225.99	47.0	1386253
O2O-PDP-C-20-80-1	10	197.43	58.2	**1153732**	206.97	59.1	1148144	197.51	66.0	1023565
O2O-PDP-C-20-80-2	10	301.19	56.2	2337252	303.19	58.3	**2338939**	301.40	59.9	2003647
O2O-PDP-C-20-80-3	10	309.55	60.9	**2422807**	312.17	63.0	2418103	309.75	47.1	2083754
O2O-PDP-C-24-96-1	10	301.54	59.3	**1636369**	301.23	57.0	1629611	301.64	66.4	1417938
O2O-PDP-C-24-96-2	10	314.04	69.6	**2789114**	313.38	71.8	2781033	314.32	66.2	2326187
O2O-PDP-C-24-96-3	10	324.49	71.9	**2886859**	317.71	70.8	2879702	324.88	36.1	2422490

Table 4.4 Performance comparison of NSGA-II and BMA on O2O-PDPTW-D instances

Set	#Solved	BMA1			BMA2			NSGA-II		
		Time(s)	#ND	*H*	Time(s)	#ND	*H*	Time(s)	#ND	*H*
O2O-PDP-D-6-48-1	10	37.10	57.4	494360.8	40.02	53.6	**499682**	37.19	55.6	478195.9
O2O-PDP-D-6-48-2	10	63.05	51.3	**865455.6**	65.85	55.6	864729.8	63.13	49.6	803757.7
O2O-PDP-D-6-48-3	10	85.64	55.6	**876407.3**	86.40	52.8	850954.7	85.76	40.7	850100.2
O2O-PDP-D-8-64-1	10	89.43	56.6	**760068.8**	98.62	54.1	756335.4	89.49	56.2	680686
O2O-PDP-D-8-64-2	10	175.74	55.8	1422790	181.41	57.9	**1424827**	175.82	57.1	1277179
O2O-PDP-D-8-64-3	10	238.16	55.8	**1631154**	238.89	55.1	1589182	238.37	37.1	1352790
O2O-PDP-D-12-96-1	10	302.26	54.5	**1687338**	300.95	59.7	1684859	302.40	70.6	1433301
O2O-PDP-D-12-96-2	10	315.36	71.7	**2792077**	311.19	73.9	2786656	315.73	50.2	2420442
O2O-PDP-D-12-96-3	10	319.91	78.4	**3068791**	318.30	76.0	3064651	320.41	30.4	2635899
O2O-PDP-D-16-128-1	10	314.53	82.6	**3091380**	311.98	84.7	3081811	314.74	74.5	2778504
O2O-PDP-D-16-128-2	10	323.82	85.5	**5125241**	324.19	84.1	5122773	324.65	42.7	4616622
O2O-PDP-D-16-128-3	10	329.98	81.0	5189661	333.42	80.2	**5190810**	331.15	27.2	4673803
O2O-PDP-D-20-160-1	10	317.82	89.4	**4149804**	321.56	90.9	4145639	318.14	75.2	3655229
O2O-PDP-D-20-160-2	10	320.95	81.7	5353188	324.35	80.4	**5355948**	322.78	24.1	5016744
O2O-PDP-D-20-160-3	10	348.99	77.3	**5627476**	345.94	77.2	5622140	350.73	21.9	5193330
O2O-PDP-D-24-192-1	10	335.39	85.7	**4535095**	317.34	84.6	4520672	336.38	43.6	4188831
O2O-PDP-D-24-192-2	10	342.53	79.6	6796357	343.67	79.2	**6862260**	345.44	21.7	6170181
O2O-PDP-D-24-192-3	10	351.83	65.7	7252149	353.99	64.6	**7252247**	355.77	20.9	6618060

Table 4.5 Comparison results for O2O-PDPTW instances in four categories

#restaurant	T_{tw}	Category A			Category B			Category C			Category D		
		#rider	*TT*	*WT*	#rider	*TT*	*WT*	#rider	*TT*	*WT*	#rider	*TT*	*WT*
6	1	1.7	315.6	89.7	2.8	478.1	197.3	3.8	701.3	399.8	7.1	1277.5	760.5
6	2	1.6	301.7	98.1	2.6	431.9	223.0	3.6	659.6	440.1	7.0	1275.2	802.6
6	3	1.6	301.8	97.7	2.6	429.0	226.7	3.6	671.9	440.4	6.9	1251.6	838.6
8	1	2.4	355.4	123.2	3.5	593.0	253.3	5.6	889.8	542.8	8.7	1611.6	1002.3
8	2	2.2	338.5	148.3	3.3	568.6	273.9	5.3	861.0	592.9	8.7	1611.9	1056.0
8	3	2.2	334.5	148.3	3.3	572.2	269.6	5.3	857.9	600.1	8.5	1590.7	1072.9
12	1	3.0	491.8	195.3	4.1	691.4	394.0	7.0	1290.4	749.6	12.3	2319.5	1478.2
12	2	2.6	430.6	234.1	3.9	655.5	431.6	6.8	1251.3	810.2	11.9	2225.0	1684.2
12	3	2.6	425.1	241.2	3.9	664.4	431.1	7.0	1274.4	804.2	11.8	2227.1	1759.8
16	1	3.7	564.0	288.8	5.5	976.2	514.7	8.6	1636.3	984.6	16.7	3156.9	1985.8
16	2	3.4	531.2	316.2	5.1	919.1	565.8	8.5	1608.5	1034.8	16.8	3175.6	2112.2
16	3	3.4	527.0	324.6	5.2	917.7	568.0	8.6	1615.3	1032.5	16.7	3180.5	2222.0
20	1	4.1	667.1	314.9	5.9	1073.9	650.4	10.7	1987.6	1195.0	20.0	3920.9	2428.5
20	2	3.7	617.7	347.3	5.8	1064.4	681.9	10.3	1920.6	1337.8	20.1	3949.9	2581.8
20	3	3.7	622.7	345.8	5.8	1065.3	712.1	10.3	1927.1	1403.7	20.0	3955.0	2628.0
24	1	4.3	792.3	394.1	6.7	1316.4	745.5	12.8	2402.5	1472.5	24.8	4849.9	3030.9
24	2	4.0	748.7	439.6	6.5	1281.7	790.1	12.5	2334.3	1640.2	24.5	4799.2	3219.8
24	3	4.0	756.2	441.0	6.7	1300.8	786.6	12.6	2365.6	1678.9	24.1	4774.3	3311.9

4.5 Conclusions

In this work, we studied the bi-objective O2O food pickup and delivery problem with time windows. Two separate objectives were considered: the total duration time related to operating cost and the total waiting time related to customer satisfaction. We applied a bi-objective memetic method which incorporates the MDLS strategy into a multi-objective evolutionary framework. The proposed approach showed good performances for instances including up to 192 orders.

Further research will focus on the extension of the method to a dynamic O2O food pickup and delivery problem to provide real-time decision making for O2O takeout companies. In addition, the combination of O2O takeout applications and sharing riders is also a possible research direction.

Chapter 5
Conclusions

5.1 Summary and management insights

The thesis investigated research problems arising from auctions in the procurement of transportation service and food pickup and delivery in O2O takeout applications. The bi-objective transportation service procurement problems were studied from the shippers' perspective whereas the bi-objective food pickup and delivery problems were studied from the O2O platforms' perspective. In both cases, efficient multi-objective optimization methods are designed and analyzed for solving these NP-hard problems.

The first problem investigated in the thesis is full truckload transportation service procurement problem with transit time (FTL-TPTT). It aimed at choosing a set of carriers for the shipper to allocate freight volume to them so as to minimize the total transportation cost and the total transit time while satisfying multiple business side constraints. An integer programming model was proposed to formulate this problem. Based on the model, a multi-objective evolutionary algorithm was applied to solve this problem. The computational results showed that the multi-objective evolutionary algorithm is efficient to explore nondominated solutions on the Pareto front. The design of the algorithm is also interesting and can be suggested to researchers who also work on the bi-objective problems that have nice structural properties in the subproblems. The algorithm can be integrated into the practical transportation service procurement systems to report fast good solutions and support decision-making for shippers.

Next, the thesis studied the transportation service procurement problem with transit time and total quantity discounts (TPTT-TQD), which generalizes the transportation service procurement problem with transit time by introducing the total quanti ty discounts. TPTT-TQD was also formulated into an integer programming problem. To obtain the Pareto front, a bi-objective branch-and-bound algorithm was designed, which including two new stronger fathoming rules, two novel bounding procedures with search region reduction, and a hybrid branching strategy. By considering the total quantity discounts in the auction, we found some insights analytically that the new auction process had a strong advantage over the original one for the shipper to improve the decision. Computational results showed that our method was capable of solving small size instances efficiently.

The third problem was the bi-objective O2O food pickup and delivery problem (O2O-PDPTW). It minimized the total duration time of all the riders and the total waiting time of all the customers simultaneously, and decided the order allocations and delivery routing plans. Due to the NP-hardness of the bi-objective problem, a bi-objective memetic method was proposed that incorporates the multi-directional local search into a multi-objective evolutionary framework. To evaluate the performance, a set of test instances was generated according to the characteristics of real-world O2O food takeout application. Computational results demonstrated the effectiveness of our approach. In addition, the correlation between time window width, number of orders, number of riders and Pareto front is also studied.

Finally, the thesis summarized the management insights observed during the creation of this thesis.

- For sea freight full container load (FCL) services, the notable carriers are only a few and an exact method is thus applicable. However, in the case of FTL freight services, where the number of carriers could be more than 30, executing the exact method would be very time-consuming; therefore, it is not a wise option. Designing a multi-objective evolutionary algorithm to approach the Pareto front efficiently will be helpful to practical shippers.
- By considering the total quantity discounts in the auction, the new auction process has a strong advantage over the original one for the shipper to improve the decision.
- In the multi-round procurement auction, if the MQCs and discounts from all the cases are brought in together, the shipper can obtain a better Pareto front, where the nondominated solutions satisfy different MQCs and achieve different discounts from different carriers.
- By choosing the broad time windows in the O2O takeout platforms, some solutions with shorter duration times can be discovered, which might be helpful to the decision makers.
- The number of riders, the total duration time and the total waiting time dramatically increase when the number of orders becomes larger. On the contrary, the number of restaurants is not a critical factor to bring in complexity. Although some orders are taken from the same restaurant, it is difficult to merge them because each order has a different time window.

5.2 Future research

A possible extension of research is to apply the bi-objective evolutionary algorithm and the bi-objective branch-and-bound algorithm to solve a class of transportation service procurement problems. Considering the business concerns, such as shipper preferences and carrier number restrictions, the solution method has to be extended to solve the new bi-objective optimization problems. In the thesis, due to practicality issues, only single shipper was considered in the transportation procurement auctions. Future work may further consider the use of collaboration transportation procurement networks to tackle multi-shipper transportation service procurement problems.

Another avenue for future work is applying the bi-objective memetic algorithm to solve other types of multi-objective PDPs, such as the bi-objective dynamic PDPs and the bi-objective PDPs under uncertainty. The objectives can also be extended to three and four objective case, e.g., the minimization of CO_2 emissions and the minimization the energy resources consumed can be developed. The trade-off between duration time, waiting time, the amount of CO_2 emissions and the energy resources consumed can be analyzed. In addition, the combination of O2O takeout applications and sharing riders is also a possible research direction.

Bibliography

[1]Ahuja, R., Magnanti, T., Orlin, J., 1993. Network flows: theory, algorithms, and applications. PrenticeHall, Upper Saddle River, NJ .

[2]Amorim, P., Günther, H.O., Almada-Lobo, B., 2012. Multi-objective integrated production and distribution planning of perishable products. International Journal of Production Economics 138, 89–101.

[3]Aneja, Y.P., Nair, K.P., 1979. Bicriteria transportation problem. Management Science 25, 73–78.

[4]Baldacci, R., Bartolini, E., Mingozzi, A., 2011. An exact algorithm for the pickup and delivery problem with time windows. Operations research 59, 414–426.

[5]Bandyopadhyay, S., Saha, S., Maulik, U., Deb, K., 2008. A simulated annealing-based multi-objective optimization algorithm: Amosa. IEEE Transactions on Evolutionary Computation 12, 269–283.

[6]Belotti, P., Soylu, B., Wiecek, M., 2013. a Branch-and-Bound Algorithm for bi-objective Mixed-Integer Programs. Optimization Online, 1–29.

[7]Bent, R., Van Hentenryck, P., 2006. A two-stage hybrid

algorithm for pickup and delivery vehicle routing problems with time windows. Computers & Operations Research 33, 875– 893.

[8]Berbeglia, G., Cordeau, J.F., Gribkovskaia, I., Laporte, G., 2007. Static pickup and delivery problems: a classification scheme and survey. Top 15, 1–31.

[9]Bérubé, J.F., Gendreau, M., Potvin, J.Y., 2009. An exact ε-constraint method for bi-objective combinatorial optimization problems: Application to the Traveling Salesman Problem with Profits. European Journal of Operational Research 194, 39–50.

[10]Boland, N., Charkhgard, H., Savelsbergh, M., 2015a. A criterion space search algorithm for bi-objective mixed integer programming: The triangle splitting method. INFORMS Journal on Computing 27, 597–618.

[11]Boland, N., Charkhgard, H., Savelsbergh, M., 2015b. A criterion space search algorithm for bi-objective integer programming: The balanced box method. INFORMS Journal on Computing 27, 735–754.

[12]Bortolini, M., Faccio, M., Ferrari, E., Gamberi, M., Pilati, F., 2016. Fresh food sustainable distribution: cost, delivery time and carbon footprint three-objective optimization. Journal of Food Engineering 174, 56–67.

[13]Bowman, V.J., 1976. On the relationship of the tchebycheff norm and the efficient frontier of multiple-criteria objectives, in: Multiple criteria decision making. Springer, pp. 76–86.

[14]Braekers, K., Hartl, R.F., Parragh, S.N., Tricoire, F., 2016. A bi-objective home care scheduling problem: Analyzing the trade-

off between costs and client inconvenience. European Journal of Operational Research 248, 428–443.

[15]Buer, T., Kopfer, H., 2014. A pareto-metaheuristic for a bi-objective winner determination problem in a combinatorial reverse auction. Computers & Operations Research 41, 208–220.

[16]Buer, T., Pankratz, G., 2010. Solving a bi-objective winner determination problem in a transportation procurement auction. Logistics Research 2, 65–78.

[17]Calvete, H.I., Galé, C., Iranzo, J.A., 2016. Meals: A multi-objective evolutionary algorithm with local search for solving the bi-objective ring star problem. European Journal of Operational Research 250, 377–388.

[18]Caplice, C., Sheffi, Y., 2003. Optimization-based procurement for transportation services. Journal of Business Logistics 24, 109–128.

[19]Cattaruzza, D., Absi, N., Feillet, D., Vidal, T., 2014. A memetic algorithm for the multi trip vehicle routing problem. European Journal of Operational Research 236, 833–848.

[20]Chalmet, L., Lemonidis, L., Elzinga, D., 1986. An algorithm for the bi-criterion integer programming problem. European Journal of Operational Research 25, 292–300.

[21]Cherkesly, M., Desaulniers, G., Irnich, S., Laporte, G., 2016. Branch-price-and-cut algorithms for the pickup and delivery problem with time windows and multiple stacks. European Journal of Operational Research 250, 782–793.

[22]Coelho, L.C., Laporte, G., 2014. Optimal joint replenishment,

delivery and inventory management policies for perishable products. Computers & Operations Research 47, 42–52.

[23]Cordeau, J.F., 2006. A branch-and-cut algorithm for the dial-a-ride problem. Operations Research 54, 573–586.

[24]Cordeau, J.F., Iori, M., Laporte, G., Salazar González, J.J., 2010. A branch-and-cut algorithm for the pickup and delivery traveling salesman problem with lifo loading. Networks 55, 46–59.

[25]Cordeau, J.F., Laporte, G., 2007. The dial-a-ride problem: models and algorithms. Annals of operations research 153, 29–46.

[26]Cordeau, J.F., Laporte, G., Ropke, S., 2008. Recent models and algorithms for one-to-one pickup and delivery problems, in: The vehicle routing problem: latest advances and new challenges. Springer, pp. 327–357.

[27]Deb, K., Pratap, A., Agarwal, S., Meyarivan, T., 2002. A fast and elitist multi-objective genetic algorithm: NSGA-II. IEEE Transactions on Evolutionary Computation 6, 182–197.

[28]Diana, M., Dessouky, M.M., 2004. A new regret insertion heuristic for solving large-scale dial-a-ride problems with time windows. Transportation Research Part B: Methodological 38, 539–557.

[29]Dolan, E.D., More, J.J., 2002. Benchmarking Optimization Software with Performance Profiles. Mathematical Programming 91, 201–213.

[30]Dorigo, M., Birattari, M., Stiitzle, T., 2006. Ant colony optimization artificial. IEEE Computational Intelligence Magazine 1, 28–39.

[31]Du, F., Evans, G.W., 2008. A bi-objective reverse logistics network analysis for post-sale service. Computers & Operations Research 35, 2617–2634.

[32]Dumas, Y., Desrosiers, J., Soumis, F., 1991. The pickup and delivery problem with time windows. European journal of operational research 54, 7–22.

[33]Ehrgott, M., Gandibleux, X., 2007. Bound sets for bi-objective combinatorial optimization problems. Computers and Operations Research 34, 2674–2694.

[34]Eshelman, L.J., Caruana, R.A., Schaffer, J.D., 1989. Biases in the crossover landscape, in: International Conference on Genetic Algorithms, pp. 10–19.

[35]França, P.M., Mendes, A., Moscato, P., 2001. A memetic algorithm for the total tardiness single machine scheduling problem. European Journal of Operational Research 132, 224– 242.

[36]Gadegaard, S., Ehrgott, M., Nielsen, L., 2016. Bi–objective branch–and–cut algorithms: Applications to the single source capacitated facility location problem.

[37]Goossens, D.R., Maas, A.J.T., Spieksma, F.C.R., van de Klundert, J.J., 2007. Exact algorithms for procurement problems under a total quantity discount structure. European Journal of Operational Research 178, 603–626.

[38]Haimes, Y.Y., Lasdon, L.S., Wismer, D.A., Haimes, Y.Y., Lasdon, L.S., Wismer, D.A., 1971. On a bicriterion formulation of the problems of integrated system identification and system optimization. Systems Man & Cybernetics IEEE Transactions on SMC-1, 296–297.

[39]Hsu, C.I., Hung, S.F., Li, H.C., 2007. Vehicle routing problem with time-windows for perishable food delivery. Journal of food engineering 80, 465–475.

[40]Hu, Q., Zhang, Z., Lim, A., 2016. Transportation service procurement problem with transit time. Transportation Research Part B: Methodological 86, 19–36.

[41]Jozefowiez, N., Laporte, G., Semet, F., 2012. A generic branch-and-cut algorithm for multi-objective optimization problems: Application to the multilabel traveling salesman problem. INFORMS Journal on Computing 24, 554–564.

[42]Kirchler, D., Calvo, R.W., 2013. A granular tabu search algorithm for the dial-a-ride problem. Transportation Research Part B: Methodological 56, 120–135.

[43]Lee, C.G., Kwon, R.H., Ma, Z., 2007. A carrier's optimal bid generation problem in combinatorial auctions for transportation procurement. Transportation Research Part E: Logistics and Transportation Review 43, 173–191.

[44]Lee, Z.J., Lee, C.Y., 2005. A hybrid search algorithm with heuristics for resource allocation problem. Information sciences 173, 155–167.

[45]Li, H., Lim, A., 2003. A metaheuristic for the pickup and delivery problem with time windows. International Journal on Artificial Intelligence Tools 12, 173–186.

[46]Lim, A., Qin, H., Xu, Z., 2012. The freight allocation problem with lane cost balancing constraint. European Journal of Operational Research 217, 26–35.

[47]Lim, A., Rodrigues, B., Xu, Z., 2008. Transportation procurement with seasonally varying shipper demand and volume guarantees. Operations research 56, 758–771.

[48]Lim, A., Wang, F., Xu, Z., 2006. A transportation problem with minimum quantity commitment. Transportation Science 40, 117–129.

[49]Lim, A., Xu, Z., 2006. The bottleneck problem with minimum quantity commitments. Naval Research Logistics (NRL) 53, 91–100.

[50]Liu, B., Wang, L., Jin, Y.H., 2007. An effective pso-based memetic algorithm for flow shop scheduling. IEEE Transactions on Systems, Man, and Cybernetics, Part B (Cybernetics) 37, 18–27.

[51]Liu, M., Lee, C.Y., Zhang, Z., Chu, C., 2016. Bi-objective optimization for the container terminal integrated planning. Transportation Research Part B: Methodological 93, 720–749.

[52]Lu, Q., Dessouky, M., 2004. An exact algorithm for the multiple vehicle pickup and delivery problem. Transportation Science 38, 503–514.

[53]Lu, Q., Dessouky, M.M., 2006. A new insertion-based construction heuristic for solving the pickup and delivery problem with time windows. European Journal of Operational Research 175, 672–687.

[54]Ma, Z., Kwon, R.H., Lee, C.G., 2010. A stochastic programming winner determination model for truckload procurement under shipment uncertainty. Transportation Research Part E: Logistics and Transportation Review 46, 49–60.

[55]Ma, Z.J., Wu, Y., Dai, Y., 2017. A combined order selection and time-dependent vehicle routing problem with time widows for

perishable product delivery. Computers & Industrial Engineering 114, 101–113.

[56]Maimaiti, M., Zhao, X., Jia, M., Ru, Y., Zhu, S., 2018. How we eat determines what we become: opportunities and challenges brought by food delivery industry in a changing world in China. European journal of clinical nutrition 72, 1282.

[57]Manerba, D., Mansini, R., 2012. An exact algorithm for the Capacitated Total Quantity Discount Problem. European Journal of Operational Research 222, 287–300.

[58]Mansini, R., Savelsbergh, M.W.P., Tocchella, B., 2012. The supplier selection problem with quantity discounts and truckload shipping. Omega 40, 445–455.

[59]Masson, R., Lehuédé, F., Péton, O., 2013. An adaptive large neighborhood search for the pickup and delivery problem with transfers. Transportation Science 47, 344–355.

[60]Mavrotas, G., Diakoulaki, D., 1998. A branch and bound algorithm for mixed zero-one multiple objective linear programming. European Journal of Operational Research 107, 530–541.

[61]Miettinen, K., 1999. Nonlinear multi-objective optimization, volume 12 of international series in operations research and management science.

[62]Molenbruch, Y., Braekers, K., Caris, A., Berghe, G.V., 2017. Multi-directional local search for a bi-objective dial-a-ride problem in patient transportation. Computers & Operations Research 77, 58–71.

[63]Moscato, P., Cotta, C., Mendes, A., 2004. Memetic algorithms, in: New optimization techniques in engineering. Springer, pp. 53–85.

[64]Nagata, Y., Bräysy, O., Dullaert, W., 2010. A penalty-based edge assembly memetic algorithm for the vehicle routing problem with time windows. Computers & operations research 37, 724–737.

[65]Nanry, W.P., Barnes, J.W., 2000. Solving the pickup and delivery problem with time windows using reactive tabu search. Transportation Research Part B: Methodological 34, 107– 121.

[66]Neri, F., Cotta, C., Moscato, P., 2012. Handbook of memetic algorithms. volume 379. Springer.

[67]Ngueveu, S.U., Prins, C., Calvo, R.W., 2010. An effective memetic algorithm for the cumulative capacitated vehicle routing problem. Computers & Operations Research 37, 1877– 1885.

[68]Pacheco, J., Caballero, R., Laguna, M., Molina, J., 2013. Bi-objective bus routing: an application to school buses in rural areas. Transportation Science 47, 397–411.

[69]Paquette, J., Cordeau, J.F., Laporte, G., Pascoal, M.M., 2013a. Combining multicriteria analysis and tabu search for dial-a-ride problems. Transportation Research Part B: Methodological 52, 1–16.

[70]Paquette, J., Cordeau, J.F., Laporte, G., Pascoal, M.M., 2013b. Combining multicriteria analysis and tabu search for dial-a-ride problems. Transportation Research Part B: Methodological 52, 1–16.

[71]Parragh, S.N., Doerner, K.F., Hartl, R.F., 2008. A survey on pickup and delivery problems. Journal fu¨r Betriebswirtschaft 58, 21–51.

[72]Parragh, S.N., Doerner, K.F., Hartl, R.F., 2010. Variable neighborhood search for the dial-aride problem. Computers & Operations Research 37, 1129–1138.

[73]Parragh, S.N., Doerner, K.F., Hartl, R.F., Gandibleux, X., 2009. A heuristic two-phase solution approach for the multi-objective dial-a-ride problem. Networks: An International Journal 54, 227–242.

[74]Parragh, S.N., Tricoire, F., 2015. Branch-and-bound for bi-objective integer programming. Optimization Online .

[75]Perugia, A., Moccia, L., Cordeau, J.F., Laporte, G., 2011. Designing a home-to-work bus service in a metropolitan area. Transportation Research Part B: Methodological 45, 1710–1726.

[76]Przybylski, A., Gandibleux, X., 2017. Multi-objective branch and bound. European Journal of Operational Research 260, 856–872.

[77]Qin, H., Luo, M., Gao, X., Lim, A., 2012. The freight allocation problem with all-units quantity-based discount: A heuristic algorithm. Omega 40, 415–423.

[78]Qu, Y., Bard, J.F., 2014. A branch-and-price-and-cut algorithm for heterogeneous pickup and delivery problems with configurable vehicle capacity. Transportation Science 49, 254– 270.

[79]Rekik, M., Mellouli, S., 2012. Reputation-based winner determination problem for combinatorial transportation procurement auctions. Journal of the Operational Research Society 63, 1400–1409.

[80]Remli, N., Rekik, M., 2013. A robust winner determination problem for combinatorial transportation auctions under uncertain shipment volumes. Transportation Research Part C: Emerging Technologies 35, 204–217.

[81]Ropke, S., Cordeau, J.F., 2009. Branch and cut and price for the pickup and delivery problem with time windows. Transportation

Science 43, 267–286.

[82]Ropke, S., Cordeau, J.F., Laporte, G., 2007. Models and branch-and-cut algorithms for pickup and delivery problems with time windows. Networks: An International Journal 49, 258–272.

[83]Ropke, S., Pisinger, D., 2006. An adaptive large neighborhood search heuristic for the pickup and delivery problem with time windows. Transportation science 40, 455–472.

[84]Savelsbergh, M.W., 1992. The vehicle routing problem with time windows: Minimizing route duration. ORSA journal on computing 4, 146–154.

[85]Sheffi, Y., 2004. Combinatorial auctions in the procurement of transportation services. Interfaces 34, 245–252.

[86]Song, B.D., Ko, Y.D., 2016. A vehicle routing problem of both refrigerated-and general-type vehicles for perishable food products delivery. Journal of Food Engineering 169, 61–71.

[87]Sourd, F., Spanjaard, O., 2008. A multi-objective branch-and-bound framework: Application to the bi-objective spanning tree problem. INFORMS Journal on Computing 20, 472–484.

[88]Stidsen, T., Andersen, K.A., Dammann, B., 2014. A branch and bound algorithm for a class of bi-objective mixed integer programs. Management Science 60, 1009–1032.

[89]Tavakkoli-Moghaddam, R., Safaei, N., Sassani, F., 2009. A memetic algorithm for the flexible flow line scheduling problem with processor blocking. Computers & Operations Research 36, 402–414.

[90]Ting, C.K., Liao, X.L., 2013. The selective pickup and delivery problem: formulation and a memetic algorithm. International Journal

of Production Economics 141, 199–211.

[91]Tricoire, F., 2012. Multi-directional local search. Computers & operations research 39, 3089–3101.

[92]Trustdata, 2018. Q1 chinese takeaway O2O industry analysis report (in Chinese). https://www.useit.com.cn/thread-19160-1-1.html.

[93]Ulungu, E.L., Teghem, J., 1995. The two phases method: An efficient procedure to solve bi-objective combinatorial optimization problems. Annals of Mathematics & Artificial Intelligence 20, 227–265.

[94]Vincent, T., Seipp, F., Ruzika, S., Przybylski, A., Gandibleux, X., 2013. Multiple objective branch and bound for mixed 0-1 linear programming: Corrections and improvements for the bi-objective case. Computers & Operations Research 40, 498–509.

[95]Wang, X., Sun, X., Dong, J., Wang, M., Ruan, J., 2017. Optimizing terminal delivery of perishable products considering customer satisfaction. Mathematical Problems in Engineering 2017.

[96]Wang, X., Wang, M., Ruan, J., Zhan, H., 2016. The multi-objective optimization for perishable food distribution route considering temporal-spatial distance. Procedia Computer Science 96, 1211–1220.

[97]Wang, Y., ying Yu, L., 2012. Optimization model of refrigerated food transportation, in: Service Systems and Service Management (ICSSSM), 2012 9th International Conference on, IEEE. pp. 220–224.

[98]Wang, Z., Tang, K., Yao, X., 2010. A memetic algorithm for multi-level redundancy allocation. IEEE Transactions on reliability 59, 754–765.

[99]Xu, S.X., Huang, G.Q., 2013. Transportation service

procurement in periodic sealed double auctions with stochastic demand and supply. Transportation Research Part B: Methodological 56, 136–160.

[100]Xu, S.X., Huang, G.Q., 2014. Efficient auctions for distributed transportation procurement. Transportation Research Part B: Methodological 65, 47–64.

[101]Xue, L., Luo, Z., Lim, A., 2016. Exact approaches for the pickup and delivery problem with loading cost. Omega 59, 131–145.

[102]Yagmahan, B., Yenisey, M.M., 2008. Ant colony optimization for multi-objective flow shop scheduling problem. Computers & Industrial Engineering 54, 411–420.

[103]Zhang, B., Yao, T., Friesz, T.L., Sun, Y., 2015a. A tractable two-stage robust winner determination model for truckload service procurement via combinatorial auctions. Transportation Research Part B: Methodological 78, 16–31.

[104]Zhang, Q., Li, H., 2007. MOEA/D: A multi-objective evolutionary algorithm based on decomposition. IEEE Transactions on Evolutionary Computation 11, 712–731.

[105]Zhang, Z., Liu, M., Lim, A., 2015b. A memetic algorithm for the patient transportation problem. Omega 54, 60–71.

[106]Zhu, Z., Xiao, J., He, S., Ji, Z., Sun, Y., 2016. A multi-objective memetic algorithm based on locality-sensitive hashing for one-to-many-to-one dynamic pickup-and-delivery problem. Information Sciences 329, 73–89.

[107]Zitzler, E., Thiele, L., 1999. Multi-objective evolutionary algorithms: a comparative case study and the strength pareto

approach. IEEE Transactions on Evolutionary Computation 3, 257–271.

[108]Zitzler, E., Thiele, L., Laumanns, M., Fonseca, C.M., Da Fonseca, V.G., 2003. Performance assessment of multi-objective optimizers: an analysis and review. IEEE Transactions on Evolutionary Computation 7, 117–132.

Acknowledgements

It took me three and a half years to finish the journey to pursuit my PhD degree. This period is full of pain and hardships, passion and harvest. Besides me, lots of people have contributed to the thesis, and I would like to express my gratitude.

Foremost, I would like to express my sincere deepest respect to my supervisor Prof.Andrew Lim for accepting me as his PhD student, and leading me into the area of operations research and management science. In 2017, he invited me to visit the National University of Singapore for half a year, where I had a wonderful and unforgettable time in my life.

I also would like to express my thanks to Dr. Qian Hu. This thesis was completed under the guidance of him. From the choice of the subject to the final completion of the subject, he has been giving me careful guidance and relentless support. His serious scientific attitude, meticulous academic spirit, the work style of excellence deeply inspired me.

Also I would like to express my thanks to Dr. Zhenzhen Zhang at National University of Singapore for the co-authorship of papers. Special gratitude goes to Dr. Zhixing Luo, Dr. Fan Liu and Dr. Caihua Chen at Nanjing university. They all provided valuable suggestions on the completion of this thesis.

My thanks also go to my friends, colleagues and companions during my study period: Dr. Ting Wang, Ms. Xue Yan, Dr. Deyan Yang, Ms. Yuan Xu, Mr. Libo Zhang, Dr. Yunlong Yu, Dr. Ke Lu, Mr. Heng Du, Mr. Haohua Li, Mr. Yuwei Wu, Ms. Yuping Jiang, Ms. Ying Liu, Mr. Ruiming Lv, Ms. Xiaohui Ding, Ms. Yijing Liang, Mr. Qingyang Wang, Mr. Xiangyu Kong, Ms. Yuyang Han, Ms. Shen Zhang and others who are not mentioned here, for all the time we spent together.

On the personal side, I would like to manifest my deep gratitude to my parents and parents-in-law, for their unconditional love and support through my life. I would like to especially thank my husband for his understanding and companionship throughout the years of my research. Finally, I would like to thank my son Andrew, who has been always my heart, my joy, my direction.

Mo Zhang

This book is the result of a co-publication agreement between China Financial and Economic Publishing House (China) and Paths International Ltd (UK).

Title: Models and Algorithms for Multi-objective Transportation Optimization Problems
Author: Zhang Mo
ISBN: 978-1-84464-690-6
Ebook ISBN: 978-1-84464-691-3

Paths International Ltd

Published in the United Kingdom
www.pathsinternational.com